"This fascinating book will teach readers to see as animals see, to be a little more visual and a little less verbal."—*Booklist*

"A lively and absorbing look at the world from animals' point of view."
—*Publishers Weekly,* starred review

"Deeply moving and fascinating."—Oliver Sacks

"*Animals in Translation* is *wonderful!* The most important book I've read in thirty years."—Patricia McConnell, author of *The Other End of the Leash*

"Two thumbs up for this thoughtful and educational compilation."
—Dr. Nicholas H. Dodman, author of *Dogs Behaving Badly*

"If one is interested in learning more about the lives and needs of animals, *Animals in Translation* is a must-read. I found it impossible to put down."
—Monty Roberts, author of *The Man Who Listens to Horses*

"Temple Grandin's insights are absolutely fascinating, groundbreaking contributions to the field of animal awareness. She is an inspiration to us all."
—Alex Shoumatoff, author of *The World Is Burning*

"In this insightful, quirky, and often funny volume, Temple Grandin takes us deep inside the minds of animals. Grandin opens new vistas that will be invaluable to anyone who cares about the creatures of the earth and sky."
—Andrew Solomon, author of *The Noonday Demon*

"Animal lovers and people lovers will be thoroughly charmed by Temple Grandin's latest book. Its sweetly simple style, chock-full of fresh and funny anecdotes, somehow delivers brilliant insights into the way animals and autistic people perceive the world."
—Candace B. Pert, Ph.D., author of *Molecules of Emotion*

"A fascinating and compelling book, filled with wisdom and insight, that lives up to its promise of decoding animal behavior."
—Elizabeth Marshall Thomas, author of *The Hidden Life of Dogs*

New York Times Bestseller • *Birmingham News* Bestseller • *Denver Post* Bestseller • *Rocky Mountain News* Bestseller • *Minneapolis Star Tribune* Bestseller • Indie Next List Selection • New Atlantic Independent Booksellers Association Bestseller • Pacific Northwest Booksellers Association Bestseller • Heartland Indie Bestseller

Praise for *Animals Make Us Human*

"Full of small fascinating details . . . A sort of instruction manual on the core emotions that make animals . . . tick, and how we can keep them happier and mentally healthier."—*Boston Globe*

"If you liked the first [book], you're going to like the second . . . As a reader I'd be curious to watch [Grandin's] mind play over other topics."
—*New York Times*

"For most readers, Grandin's book will contain at least a few surprises . . . But for those of us whose connection to animals is through the heart, her words ring every bit as true."—*Christian Science Monitor*

"Chock-full of astounding insights . . . [with] heaps of suggestions for concrete ways pet owners, livestock handlers and others can improve living conditions for animals . . . A well-written, down-to-earth look into the lives of lots of animals."—*Rocky Mountain News*

"Thought-provoking."—*Milwaukee Journal Sentinel*

"Grandin . . . has become not just a voice for the world of autism, but also an interpreter between animal and human worlds. There is simply no one like her."—*Philadelphia Inquirer*

"[Grandin's] chapters on dogs and cats will be the most immediately rewarding . . . but there's a world of insight to be gained from her work on farm animals as well as more exotic zoo animals. Grandin shows a startling tenderness . . . Anybody who thinks autistic people lack empathy should read *Animals Make Us Human*."—*Time*

"This volume is loaded with small revelations . . . [Grandin's] insights are invaluable."—*The Week*

"Grandin continues to enlighten about animal instincts and emotions. Plain-spoken and good humored, Dr. Grandin has autism and that gives her a unique perspective. She is a voice for each animal she discusses."
—*Morning Call* (Pennsylvania)

"Grandin expresses smart ideas through accessible language."
—OregonLive.com

"Packed with fascinating insights, unexpected observations and a wealth of how-to tips, Grandin's peppy work ably challenges assumptions about what makes animals happy."—*Publishers Weekly*, starred review

"Extraordinary."—*BookPage*

"Thought-provoking . . . Readers will be able to glean new perspectives about animal welfare."—*Library Journal*

"A specially gifted 'animal translator' shares fascinating insights and observations on how we can treat other creatures in ways that are more truly humane . . . An engaging nature book that spells out how we can make animals happy."—*Barnes & Noble Recommends*

"Sure to become a classic in the field of human/animal relationships."
—Patricia McConnell, author of *The Other End of the Leash*

"So informative and compelling that most readers will cheer [Grandin] on in her ongoing battle to ensure that animals have decent, satisfying lives and painless, fear-free deaths."—*Globe and Mail*

"Grandin's affable and endearing writing style is peppered with a surprising amount of humour . . . [She] takes us on a journey of self-realization that animals are indeed us."—*Winnipeg Free Press*

Praise for *Animals in Translation*

"Inspiring . . . Crammed with facts and anecdotes about Temple Grandin's favorite subject: the senses, brains, emotions, and amazing talents of animals."—*New York Times Book Review*

"Full of heart, soul and crackling intelligence."—*People*

"Neurology has Oliver Sacks, nature has Annie Dillard, and the lucky an world has Grandin, a master intermediary between humans and our fe beasts . . . At once hilarious, fascinating, and just plain weird, *Animals* is of those rare books that elicits a 'wow' on almost every page. A."
—*Entertainment Wee*

"[Grandin's] unique perspective is like one of those fairy tales that reveal magical kingdom lurking in a cranny of your kitchen. Except that porta isn't your cupboard, it's your cat or dog."—*Boston Globe*

"[*Animals in Translation*] is suffused with a love of all creatures great and small, and full of helpful hints for raising pets."—*Parade*

"Grandin's focus in *Animals in Translation* is not on all the 'normal' things autistics and animals can't do but on the unexpected, extraordinary, invaluable things they can."—*O, the Oprah Magazine*

"At times, it is difficult to work out whether this is a book about animal behavior with insight from autism, or a book about autism that uses animal behavior to explain what it is like to be autistic. A major achievement of the book is that it is both."—*Nature*

"Earnest, humble and compassionate."—*Cleveland Plain Dealer*

"Provocative, clear and funny, *Animals in Translation* calls into question who and what is 'normal.'"—*Minneapolis Star Tribune*

"Not since Jane Goodall's research on the chimpanzee's use of tools has there been a book that so successfully challenges our definitions of what is human and what is animal."—*BookPage*

"Entertaining and insightful."—*Rocky Mountain News*

"Grandin's new book draws a fascinating parallel between autism and the way animals process the world around them."—*Newsday*

"Grandin reminds us, as almost no one else has been able to do, that humans are not humans without animals."—*Discover*

Also by Temple Grandin and Catherine Johnson

ANIMALS IN TRANSLATION: Using the Mysteries of Autism to Decode Animal Behavior

Also by Temple Grandin

EMERGENCE: Labeled Autistic (*with Margaret M. Scariano*)

THINKING IN PICTURES: And Other Reports from My Life with Autism

GENETICS AND THE BEHAVIOR OF DOMESTIC ANIMALS

LIVESTOCK HANDLING AND TRANSPORT

DEVELOPING TALENTS (*with Kate Duffey*)

HUMANE LIVESTOCK HANDLING (*with Mark Deesing*)

Also by Temple Grandin and Sean Barron

UNWRITTEN RULES OF SOCIAL RELATIONSHIPS

Also by Catherine Johnson

SHADOW SYNDROMES: The Mild Forms of Major Mental Disorders That Sabotage Us (*with John J. Ratey*)

LUCKY IN LOVE: The Secrets of Happy Couples and How Their Marriages Thrive

WHEN TO SAY GOODBYE TO YOUR THERAPIST

ANIMALS

MAKE US HUMAN

Creating the Best Life for Animals

TEMPLE GRANDIN
and Catherine Johnson

Mariner Books • Houghton Mifflin Harcourt • BOSTON • NEW YORK

First Mariner Books edition 2010

www.hmhbooks.com

Library of Congress Cataloging-in-Publication Data

Grandin, Temple.
Animals make us human : creating the best life for animals /
Temple Grandin & Catherine Johnson.
p. cm.
1. Emotions in animals. 2. Animal behavior.
I. Johnson, Catherine, date II. Title.
QL785.27.G73 2009 636.08'32—dc22 2008034892
ISBN 978-0-15-101489-7
ISBN 978-0-547-24823-3 (pbk.)

Printed in the United States of America

Book design by Robert Overholtzer

DOM 10 9 8 7 6 5 4 3 2 1

Contents

1 What Do Animals Need?

WHAT DOES AN ANIMAL NEED to have a good life?

I don't mean a good life physically. We know a lot about what kind of food, water, exercise, and veterinary care animals need to grow well and be healthy.

I mean a good mental life.

What does an animal need to be happy?

The animal welfare movement has been thinking about animals' mental welfare at least since the 1960s. That's when the British government commissioned the Brambell Report on intensive animal production. *Intensive animal production* means very big farms raising large numbers of animals for slaughter or egg production in very small spaces compared to traditional farms. The Brambell committee listed the five freedoms animals should have. The first three freedoms are about physical welfare, and the last two are about mental welfare:

- freedom from hunger and thirst
- freedom from discomfort
- freedom from pain, injury, or disease
- freedom to express normal behavior
- freedom from fear and distress

Freedom is a confusing guide for people trying to give animals a good life. Even freedom from fear, which sounds straightforward, isn't simple or obvious. For example, zookeepers and farmers usually assume that as long as a prey species animal doesn't have any predators around, it can't be afraid. But that's not the way fear works inside the brain. If you felt fear only when you are face-to-face with the animal that's going to kill you and eat you, that would be too late. Prey species animals feel afraid when they're out in the open and exposed to potential predators. For example, a hen has to have a place to hide when she lays her eggs. It doesn't matter that she's laying her eggs on a commercial farm inside a barn that no fox will ever get into. The hen has evolved to hide when she lays her eggs. Hiding is what gives her freedom from fear, not living in a barn that keeps the foxes out. I'll talk more about this in my chapter on chickens.

The freedom to express normal behavior is even more complicated and hard to apply in the real world. In many cases, it's impossible to give a domestic or captive animal the freedom to express a normal behavior. For a dog, normal behavior is to roam many miles a day, which is illegal in most towns. Even if it's not illegal, it's dangerous. So you have to figure out substitute behaviors that keep your dog happy and stimulated.

In other cases, we don't know how to create the right living conditions because we don't know enough about what the normal behavior of a particular animal is. Cheetahs are a good example. Zookeepers tried to breed cheetahs for years with almost no success. That's a common problem in zoos. Breeding is one of the most basic and normal behaviors there is. There wouldn't be any animals or people without it. But a lot of animals living in captivity don't mate successfully because there's something wrong with their living conditions that stops them from acting naturally. The cheetah-breeding problem was finally solved in 1994, when a study of cheetahs on the Serengeti Plains came out and everyone realized male and female cheetahs didn't live together in the wild the way they did in zoos. When zoos separated the female cheetahs from the males, they turned out to be easy to breed in captivity.[1]

Animal distress is even more mysterious. What is distress in an animal? Is it anger? Is it loneliness? Is it boredom? Is boredom a feeling? And how can you tell if an animal is lonely or bored?

Although a lot of good work has been done on mental welfare for animals, it's hard for pet owners, farmers, ranchers, and zookeepers to use it because they don't have clear guidelines. Right now, when a zoo wants to improve welfare, what usually happens is that the staff tries everything they can think of that they have the money and the personnel to implement. Mostly they focus on the animal's behavior and try to get it acting as naturally as possible.

I believe that the best way to create good living conditions for any animal, whether it's a captive animal living in a zoo, a farm animal, or a pet, is to base animal welfare programs on the core emotion systems in the brain. My theory is that the environment animals live in should activate their positive emotions as much as possible, and not activate their negative emotions any more than necessary. If we get the animal's emotions right, we will have fewer problem behaviors.

That might sound like a radical statement, but some of the research in neuroscience has been showing that emotions drive behavior, and my own thirty-five years of experience working with animals have shown me that this is true. Emotions come first. You have to go back to the brain to understand animal welfare.

Of course, usually — though not always — the more freedom you give an animal to act naturally, the better, because normal behaviors evolved to satisfy the core emotions. When a hen hides to lay her eggs, the hiding behavior turns off fear. But if you can't give an animal the freedom to act naturally, then you should think about how to satisfy the emotion that motivates the behavior by giving the animal other things to do. Focus on the emotion, not the behavior.

So far, research in animal behavior agrees with the neuroscience research on emotions. A really good study on whether animals have purely behavioral needs was done with gerbils. Gerbils love to dig and tunnel, and a lot of them develop a corner-digging stereotypy when they're around thirty days old. A *stereotypy* is an abnormal repetitive behavior (ARB for short), such as a lion or tiger pacing back and forth in its cage for hours on end. Pets and farm animals can develop stereotypies, too. Stereotypies are defined as abnormal behaviors that are repetitive, invariant (lions always pace the exact same path in their cages), and seemingly pointless.

An adult gerbil spends up to 30 percent of its "active time" doing stereotypic digging in the corner of its cage. That would never happen in nature, and many researchers have hypothesized that the reason captive gerbils develop stereotypic digging is that they have a biological need to dig that they can't express inside a cage.

On the other hand, in nature gerbils don't dig just to be digging. They dig to create underground tunnels and nests. Once they've hollowed out their underground home, they stop digging. Maybe what the gerbil needs is the result of the digging, not the behavior itself. A Swiss psychologist named Christoph Wiedenmayer set up an experiment to find out. He put one set of baby gerbils in a cage with dry sand they could dig in, and another set in a cage with a predug burrow system but nothing soft to dig in. The gerbils in the sand-filled box developed digging stereotypies right away, whereas none of the gerbils in the cage with the burrows did.[2]

That shows that the motivation for a gerbil's digging stereotypy is a need to hide inside a sheltered space, not a need to dig.

The gerbil needs the emotion of feeling safe, not the action of digging. Animals don't have purely behavioral needs, and if an animal expresses a normal behavior in an abnormal environment, its welfare may be poor. A gerbil that spends 30 percent of its time digging without being able to make a tunnel does not have good welfare.

The Blue-Ribbon Emotions

All animals and people have the same core emotion systems in the brain. Most pet owners probably already believe this, but I find that a lot of executives, plant managers, and even some veterinarians and researchers still don't believe that animals have emotions. The first thing I tell them is that the same psychiatric medications, such as Prozac, that work for humans also work for animals.[3] Unless you are an expert, when you dissect a pig's brain it's difficult to tell the difference between the lower-down parts of the animal's brain and the lower-down parts of a human brain.[4] Human beings have a much bigger neocortex, but the core emotions aren't located in the neocortex. They're in the lower-down part of the brain.

When people are suffering mentally, they want to feel better — they want to stop having bad emotions and start having good emotions. That's the right goal with animals, too.

Dr. Jaak Panksepp, a neuroscientist at Washington State University who wrote the book *Affective Neuroscience* and is one of the most important researchers in the field, calls the core emotion systems the "blue-ribbon emotions," because they "generate well-organized behavior sequences that can be evoked by localized electrical stimulation of the brain."[5] This means that when you stimulate the brain systems for one of the core emotions, you always get the same behaviors from the animal. If you stimulate the anger system, the animal snarls and bites. If you stimulate the fear system, the animal freezes or runs away. Electrodes in the social attachment system cause the animal to make separation calls, and electrodes in the "SEEKING" system make the animal start moving forward, sniffing, and exploring its environment. When you stimulate these parts of the brain in people, they don't snarl and bite, but they report the same emotions animals show.

People and animals (and possibly birds) are born with these emotions — they don't learn them from their mothers or from the environment — and neuroscientists know a fair amount about how they work inside the brain.

Here is a quick rundown of the four blue-ribbon emotion systems, which Jaak always writes in all caps:

SEEKING. Dr. Panksepp says SEEKING is "the basic impulse to search, investigate, and make sense of the environment." SEEKING is a combination of emotions people usually think of as being different: wanting something really good, looking forward to getting something really good, and curiosity, which most people probably don't think of as being an emotion at all.[6]

The wanting part of SEEKING gives you the energy to go after your goals, which can be anything from food, shelter, and sex to knowledge, a new car, or fame and fortune. When a cat stalks a mouse, its actions are driven by the SEEKING system.

The looking-forward-to part of SEEKING is the Christmas emotion. When kids see all the presents under the Christmas tree, their SEEKING system goes into overdrive.

Curiosity is related to novelty. I think the orienting response is the first stage of SEEKING because it is attracted to novelty. When a deer or a dog hears a strange noise, he turns his head, looks, and pauses. During the pause, the animal decides, Do I keep SEEKING, run away in fear, or attack? New things stimulate the curiosity part of the SEEKING system. Even when people are curious about something familiar — like behaviorists being curious about animals, for instance — they can only be curious about some aspect they don't understand. They are

SEEKING an explanation that they don't have yet. SEEKING is always about something you don't have yet, whether it's food and shelter or Christmas presents or a way to understand animal welfare.

SEEKING is a very pleasurable emotion. If you implant electrodes into the SEEKING system of an animal's brain, it will press a lever to turn the current on. Animals like to self-stimulate the SEEKING system so much that for a long time researchers thought the SEEKING system was the brain's "pleasure center," and some people still talk about it that way.[7] But the pleasure people feel when their SEEKING system is stimulated is the pleasure of looking forward to something good, not the pleasure of having something good.[8]

SEEKING might be a kind of master emotion. Jaak Panksepp says that SEEKING could be a "generalized platform for the expression of many of the basic emotional processes . . . It is the one system that helps animals anticipate all types of rewards."[9] It's possible the SEEKING system helps you anticipate bad things, too. There is new research showing that one area in the nucleus acumbens, which is part of the SEEKING system, responds to negative stimuli the animal is afraid of.[10] The SEEKING system might turn out to be an all-purpose emotion engine that produces both positive and negative motivations to approach or to avoid. But until researchers learn more, SEEKING means the positive emotions of wanting, looking forward to, or being curious about something, and that's the way I will be using the term in this book. SEEKING feels good.

RAGE: Dr. Panksepp believes that the core emotion of RAGE evolved from the experience of being captured and held immobile by a predator. Stimulation of subcortical brain areas causes an animal to go into a rage.[11] RAGE gives a captured animal the explosive energy it needs to struggle violently and maybe shock the predator into loosening its grip long enough that the captured animal can get away. The RAGE feeling starts at birth — if you hold a human baby's arms to his sides, he will become furiously angry.

Frustration is a mild form of RAGE that is sparked by mental restraint when you can't do something you're trying to do. That's why you feel mild anger when you can't unscrew a tight lid from a jar or when you can't solve a math problem. In one case the action of opening the jar has been restrained, and in the other the mental action of solving the math problem has been restrained. Frustration from mental restraint evolved out of RAGE from physical restraint.

We should assume that some captive animals feel frustrated being locked up inside enclosures, barns, apartments and houses, yards, and cages, because being locked up is a form of restraint no matter how nice the environment is. Many captive animals try to escape as soon as they have an opportunity. That was something my dissertation adviser at the University of Illinois, Bill Greenough, used to talk about. Bill used to say that maybe when we created enriched environments for laboratory animals we were just creating an enlightened San Quentin prison. I think he was right.

FEAR: The FEAR system doesn't need a lot of explanation. Animals and humans feel FEAR when their survival is threatened in any way, from the physical to the mental and social.[12] The FEAR circuits in the subcortex of the brain have been fully mapped. Destruction of the amygdala, the brain's fear center, turns off fear.[13] The core emotion of FEAR motivated the gerbils I mentioned before to dig, because in the wild gerbils who did not dig tunnels were eaten by predators.

PANIC: PANIC is Jaak's word for the social attachment system. All baby animals and humans cry when their mothers leave, and an isolated baby whose mother does not come back is likely to become depressed and die. The PANIC system probably evolved from physical pain. When you stimulate the part of an animal's brain that regulates physical pain, the animal makes separation cries. Opioids are even more effective at treating social pain than they are at treating physical pain. Jaak says that's probably why people say it "hurts" to lose someone they love.

Dr. Panksepp also writes about three other positive emotion systems researchers don't know as much about, and that don't necessarily run through an animal's entire life. He calls these three emotions "more sophisticated special-purpose socioemotional systems that are engaged at appropriate times in the lives of all mammals."

LUST: LUST means sex and sexual desire.

CARE: CARE is Dr. Panksepp's term for maternal love and caretaking.[14]

PLAY: PLAY is the brain system that produces the kind of roughhousing play all young animals and humans do at the same stage in their development. The parts of the brain that motivate PLAY are in the subcortex.[15] No one understands the nature of playing or the PLAY system in the brain well yet, although we do know that play behavior is probably a sign of good welfare, because an animal that's depressed, frightened, or angry doesn't play. The PLAY system produces feelings of joy.

Taken together, these seven emotions — especially the first four — explain why some environments are good for animals (and people) and others are bad. In a good environment you have healthy brain development and few behavior issues.

Pigs in Disneyland

The Brambell Report said animals should be free to express normal behaviors, but it didn't say animals have to have natural environments. For as long as I've been working in the field of animal behavior and welfare, "enriched environments" have been the main approach to giving animals a good emotional life.

The idea that animals are happier in enriched environments first came from research psychologists working with lab rats. In the 1940s, Donald Hebb, a Canadian psychologist, raised some young rats in his house instead of in a laboratory cage. Later on, when he tested them, they had higher intelligence and better problem-solving abilities than the rats that grew up in cages.

Twenty years later, in the 1960s, a research psychologist named Mark Rosenzweig was the second major researcher to study lab rats in enriched environments.[16] No one in the general public has ever heard of him even though he showed that an adult brain could grow new cells, a finding that went totally against everything neuroscientists believed. Dr. Rosenzweig's enriched adult rats had an 8 percent increase in thickness of the cerebral cortex.[17] That was an amazing finding, but nobody picked up on the idea that the brain could be plastic (could grow and change) in adult rats as well as juveniles.

Bill Greenough's experiments in the late 1960s and 1970s raising baby rats in stimulating environments were the studies that became famous. Bill raised one group of rats in a standard plastic laboratory cage with shavings on the floor. The other group lived in an enriched environment filled with lots of toys and old wood boards. He brought in new toys every day and changed the position of the boards, so the enriched environment also included a lot of novelty and change. When he looked at the brains, he found that the rats in the enriched environment had greater dendritic growth in their visual cortex.[18] Dendrites are tiny little threads that branch out from brain cells and conduct electrical impulses into the cell body. Rats living in stimulating environments had more brain growth.

Bill's work had a huge effect on me, and I think he influenced the whole field of animal welfare, because researchers have been studying barren and enriched environments for thirty years now. I went to the University of Illinois in 1981 to work with Bill because of that study.

When I sent in my application, I was especially concerned about the way farms were treating their pigs. There was a lot of controversy, which is still going on today, about the sow stalls where mama pigs were kept locked up for their whole pregnancy. The sow stalls were so narrow the pigs didn't even have enough room to turn around. I thought that maybe if I duplicated Bill's rat research in pigs I would have a biological test researchers could use to prove that barren environments are bad for pigs. I would be able to show that pigs raised on hard plastic floors they couldn't root in had fewer dendrites than pigs raised in nice straw-bedded pens.

So, for my dissertation research, I copied Bill's enriched rats experiment using young pigs. Twelve of my piglets lived in six baby pens with perforated plastic floors and nothing much to do. The other twelve lived in a Disneyland for pigs with lots of straw to root in and toys to play with: plastic balls, old telephone books they could rip up, boards, and a metal pipe they could roll around the floor. Every day I was putting new things in and taking old things out. New things were the key. The pigs loved fresh, new straw, which they found very interesting. The old straw was boring. You would think straw is straw, but it isn't. New straw was exciting; old straw wasn't.

My hypothesis was that the brains of the Disneyland pigs would show more dendritic growth than the brains of the barren-environment pigs. Back then the only way to compare neurons from one brain to another was to spend hours and hours staring into a microscope and drawing the cells by hand, which I did. I looked at two parts of the pigs' cortex: the visual cortex, which was where Bill's enriched rats had extra dendritic growth, and the somatosensory cortex, which receives information from the pig's snout.

When I finally got done, I realized the Disneyland pigs didn't have any greater dendritic growth at all. I was even more surprised to find out that my barren-environment pigs did have greater growth. Also, my barren-environment pigs had their extra growth in the somatosensory cortex, not the visual cortex where Bill's rats had shown extra growth.[19] My experiment totally contradicted Bill's. My enriched pigs didn't have greater brain growth, and the part of the brain where my understimulated pigs did have greater growth was different from the part where Bill's enriched rats had theirs.

When I told Bill about my results he said, "Oh, s***."

He thought I must have made a mistake, so I had to do the whole experiment over again. This time I installed a battery of security cameras trained on the pigs so I could see what they were doing when I wasn't around.

I already knew my barren-environment pigs had to be different from my Disneyland pigs, because they were so hyper. I'd go to clean the pens and they'd bite the hose over and over again and get in the way; they wouldn't stay away from me. That was from the environmental deprivation, which makes animals hyperactive. When the pigs saw the water hose, their SEEKING system went into overdrive.

I found out from watching the videotapes that they were hyper at night, too. All night long they were rubbing their noses into each other and into the floor, and they were going crazy manipulating the nipple waterer, which is a water pipe with a nipple on the end. All this activity was going on while the Disneyland pigs were sleeping.

When I looked at the brains under the microscope, I found the same thing I found the first time. The barren-environment pigs had greater dendritic growth than the Disneyland pigs, and the greater dendritic growth was in the somatosensory cortex, not the visual cortex.

Bill wasn't happy about my second experiment, either.

Trying to figure it out, I got to thinking that maybe what makes dendrites grow isn't the environment. What makes dendrites grow are the animal's behaviors and actions in its environment. Bill Greenough created a visually complex environment for his rats. There was a lot to look at. But my barren-environment piglets had been doing a lot, not seeing a lot. They'd been constantly using their noses to prod and poke each other and the waterer. Greater use of a body part led to greater dendritic growth in the part of the brain that received input from that body part. I think the lack of stimulation revved up their SEEKING system, because when I cleaned their feeders the pigs were so starved for stimulation that they intensely rooted and chewed at my hands. My Disneyland pigs were much less interested in feeder cleaning because they had plenty of fresh straw and toys to occupy their SEEKING system.

Everyone who read Bill Greenough's studies, including me, automatically assumed that increased dendritic growth was a good thing. But after I saw how my pigs were acting at night when they should have been sleeping, I started to think there can be increased dendritic growth that was abnormal and bad.

Bill didn't agree, but that's what neuroscientists believe today. You can have too little brain growth and you can have too much growth. Both things can be pathological. My barren-environment pigs probably had abnormal overgrowth of the dendrites in the somatosensory cortex. This is where my belief came that it is so important to satisfy the SEEKING system to prevent abnormal brain development.

What Makes an Environment Stimulating?

I didn't come out of graduate school with a biological test for animal welfare, and we still don't have one today. The only guide people have to judge whether an environment is good for an animal is the animal's behavior, which gives us insight into its emotion. But that raises quite a few questions. For one, we don't necessarily know how a captive or domestic animal with good mental welfare should behave, and some animals even hide the fact that their welfare is very poor. Prey species animals such as cattle and sheep hide their pain when they know they are being watched so that predators cannot detect their weakness. When nobody is around they may be lying down and moaning. Another problem with using the animal's behavior to judge its mental welfare is that captive and domestic animals aren't free to act the way they would act in the wild. For example, a normal, healthy animal can mate successfully, so if you have an animal that can't or won't mate, that's a red flag. But if a captive animal never has an opportunity to mate, there's no way to tell whether it would if it had the chance.

Probably for reasons like these, animal welfare researchers have ended up focusing on abnormal repetitive behaviors — stereotypies — to judge animal well-being. Stereotypies are extremely common, easy to see, and definitely abnormal in humans, although both people and animals in certain high-tension moments do have normal stereotypies. If you watch a tennis match, you'll see lots of them. Roger Federer has a racket-twirling stereotypy, and Maria Sharapova has a little repetitive dance she does while she's waiting for her opponent to serve. I call these "burst" stereotypies, because they don't last long. Animals do lots of burst stereotypies. Pigs go crazy bar chewing and bar biting at feeding time. Animals living in the wild also have some burst stereotypies. Polar bears are notorious pacers and figure-eight swimmers in captivity, and they've been observed doing "transient pacing" in the wild.

Burst stereotypies are probably always normal, so I don't worry about them. The stereotypies I worry about are the continuous stereotypies, the ones that go on for hours. Really intense stereotypies — stereotypies an animal spends hours a day doing — almost never occur in the wild, and they almost always do occur in humans with disorders such as schizophrenia and autism. Normal children raised in isolation also have stereotypies. One study of adopted Romanian orphans in Canada found that 84 percent of them had stereotypies. A lot of them rocked back and forth on their hands and knees inside their cribs; other babies stood up, held on to the sides of the cribs, and shifted back and forth from one foot to the other.

One-fourth of the children had self-injurious behavior, or SIB, as well. Self-injurious behavior means the children deliberately injured themselves the way some autistic children do: biting their hands, banging their heads against the wall, or slapping themselves in the face and head. Captive animals can have SIBs, especially primates. Ten to 15 percent of rhesus monkeys living alone in a cage develop self-biting, head banging, and self-slapping.

You never see ARBs or SIBs that severe in the wild. So, when you see them in captivity, that means something is wrong.

85 Million Animals

Georgia Mason and Jeffrey Rushen at the University of Guelph and Agri-Food Canada estimate that over 85 million farm, laboratory, and zoo animals and pets worldwide have stereotypies, including 91.5 percent of all pigs, 82.6 percent of poultry, 50 percent of lab mice, 80 percent of American minks living on fur farms (these are breeding females), and 18.4 percent of horses.[20]

That's a lot of stereotypies, and researchers are still trying to come up with the best way to classify the different types of stereotypy. Georgia Mason groups the most common kinds of ARBs this way:

- Pacing-type ARBs — pacing and other similar actions, such as circuit swimming, where a bear or a seal swims the same circuit around its pool over and over again. Over 80 percent of stereotyping carnivores pace, either back and forth or in a figure-eight pattern.
- Oral ARBs — bar and fence chewing, obsessive object licking, tongue rolling, and so on. Oral stereotypies are

common in all grazing animals, because that's what they do all day. They graze.

- Other ARBs — rocking, repetitive jumping, and so on, or "non-locomotory body movements."

The zoo animals I call the "big pretty animals" — the big predators such as the lions, tigers, and bears — pace. Ungulates, which are the hoofed animals — horses, cows, rhinoceroses, pigs, zebras, llamas — do stereotypies with their mouths. Most of the other animals, including primates and lab rats, develop movement stereotypies in the third category. In human disorders such as autism, the abnormal behavior is usually in the first or third category.

One of the most extreme cases of stereotypy I've ever seen was in a female wolf I saw at a wolf shelter. The wolf's name was Luna. Some crazy lady had been raising wolves in her yard, where she kept them all tied up to trees. No social roaming animal can be tied up all the time; keeping wolves or dogs tied up like that is cruel. They need to travel around and have lots of free social contact with other wolves and dogs. What that lady did was terrible.

The shelter people had rescued all the wolves and built really nice enclosures for them, one hundred feet long, thirty feet wide, and full of trees. They built six pens and put two wolves to a pen, which is fine. Wolf families are generally pretty small, maybe around seven or eight animals, so two wolves to a pen gave each wolf another wolf to socialize with, without the shelter risking putting together a lot of incompatible individuals that might get into fights.

Probably about half of all the wolves were pacers when they first got to the pens, but some of them were in worse shape than others. Luna and her pen mate were both pacing. The pen mate, though, would respond to changes in the environment. When you walked into the pen she'd look up and see you, or if a truck drove by she'd stop and look at it. If you stood in front of her while she was pacing, she'd notice you were there and take another path.

Luna was completely out of it. She was a beautiful wolf, with a gorgeous coat, and her mouth was in the relaxed "smile" position. But she acted the way some young autistic children do; she was in her own little world. You'd walk into the pen and she wouldn't be aware that you were there, and she didn't react to trucks driving by. She had paced so much she'd worn a path into the ground.

There was a log by Luna's path, so I sat down on it with my student Lily and we put our toes on the edge of Luna's path in the ground. Luna just paced by our toes like they weren't even there.

Then I stretched my leg out across her path. Luna jumped over my leg, but not in a normal way. She dropped her toes the way I've seen autistic kids do and scuffed them on my leg as she went over.

I don't know why toe dropping happens, but my own shoes were always scuffed on the top of the toe when I was a child. No other children had scuffs on top of their shoes, just me. Being autistic, I had a lot of stereotypies, too.

Next I put my other leg out, and she did the same thing. She put her toes down and scuffed them on my legs when she jumped over.

Then Lily put one leg out and the same thing happened. Luna jumped over all three of our legs without acting like they were there, and she scuffed her toes. Lily put her other leg out, so now there were four legs in the path. Luna jumped and scuffed again.

I wanted to see if there was any way to get Luna to notice that there were two human beings blocking her path, so I put my hand out about eight inches above my leg, like a low wall. Luna jumped the "wall" very badly, bashed her foot on my hand, and kept on going as if Lily and I weren't there. I raised my hand to eighteen inches above my leg, and this time Luna smashed into my hand with her chest and scuffed all four of our legs with her toes. The shelter lady told me that another woman who worked there had stood in front of Luna once, blocking her path, and Luna knocked her over. Ran right over her. Luna was like a robot, or a wolf zombie. She just kept pacing back and forth, back and forth, and nothing could catch her attention or change her path.

A Shock

When I first started writing this book, I thought that you could use stereotypies as a test of animal welfare. If a captive animal is stereotyping, that means it is suffering. The reason I thought this is that I've spent a lot of time around high-strung, nervous horses that have more stereotypies than calm horses. Also, I had stereotypies myself when I was little, and I had a lot of problems then. Repetitive behavior calmed me down when my overly sensitive nervous system was bombarded by sounds that hurt my ears.

But just a few weeks after I started to read the most recent research on stereotypies and barren environments, I found a group of studies on mink stereotypies that blew my mind. Farmed minks are high-activity animals that live in horrible, small cages. Anyone would expect them to have a lot of stereotypies, living in that tiny space, but 25 percent of the minks in the study — all breeding females — didn't have any stereotypies at all. They were not living in a good environment, but they didn't have stereotypies and they were breeding well.

That part didn't surprise me because there is a huge variability in stereotypies between different individual animals. I saw that with my pigs. The shock came when I read the results for the

75 percent of minks that *were* stereotyping. It was the opposite of everything I had always believed. The 75 percent of minks that had stereotypies were calmer and less fearful than the 25 percent that didn't.[21] They weren't out of it like Luna, either. When the experimenters pushed a stick a little way inside their cage, the stereotyping minks explored it, but the rest of the minks either attacked the stick violently or ran away. An animal that explores a novel object put inside its cage has better welfare than an animal that is terrified or enraged. The stereotyping minks had better welfare than the minks that didn't have stereotypies.

When I first read this, I was like Bill Greenough with the pig results — "Oh, s***! Oh, s***!" All I could keep thinking about was, "How do I reconcile these minks with everything else I know?" I was also freaked out because I knew there would be some people who would use the studies to say it's acceptable to keep minks in these horrible cages because the stereotyping minks are calm.

Then I went through all the new research on stereotypies and realized my mistake. I was used to seeing stereotypies in high-fear Arab horses and autistic children. So I associated all stereotypies with fear and anxiety. But the most recent research on stereotypies showed me that wasn't the whole story. Yes, stereotypies are abnormal, but you can't automatically assume that an animal that is stereotyping has poor welfare right at that moment or that an animal that is not stereotyping has good welfare. An animal that is stereotyping might have better welfare than an animal that isn't. Abnormal repetitive behavior means one of three things:

- The animal is suffering now.
- The animal was suffering sometime in the past but isn't suffering now. A barren environment caused my pigs to start doing stereotypies. I think this caused extra, abnormal

dendrites to grow. Even when the pigs were moved to a better environment, stereotypies tended to persist thanks to those extra dendrites.

- The animal's current welfare may not be great, but the animal is in better shape than other animals in the same barren facility that aren't stereotyping. A stereotyping animal in a bad environment may be soothing or stimulating itself, whereas the nonstereotyping animal may have just given up and become totally withdrawn and depressed. In a bad environment, the pacing animals have better welfare.

I would put Luna the wolf in the second category. Luna had good living conditions at the shelter, but she still had some of the worst stereotypies I've ever seen in a canine. I think stereotypies can have different motivators that are based on the core emotions. Fear may be the driver in some cases, but the minks were probably motivated by the SEEKING system. Since there is nothing to seek in a barren cage, they paced. When I replayed the memories of my childhood stereotypies, I realized that they were initially motivated by fear so I could escape from sounds that hurt my ears. I studied all the reflections on grains of sand that I dribbled through my hand, and I shut out the world around me. My SEEKING system had now kicked in, and I studied details that most people would ignore.

The reason Luna's pacing was so extreme is probably that she was born and raised in captivity. That's one of the most interesting findings from the research on animal stereotypies: wild-caught adult animals — animals that were born and grew up in the wild before being captured — have fewer stereotypies than animals raised in captivity.[22] Most people would think that animals captured in the wild and put in a zoo would be pacing or bar biting like crazy because it's horribly stressful to remove wild animals from their natural habitat and transport them to zoos, and it should never be done. But it's the other way around. Animals born in captivity have more stereotypies than animals born in the wild.

The reason wild-caught animals stereotype less than animals born and raised in captivity is probably that wild-caught animals were living in a rich, natural environment when they were young and their brains were developing. Many animals born in captivity were raised in barren environments like the Romanian orphans. Luna was probably a deprived animal with a scar on her brain that caused her pacing to be worse.

That explains the pet tiger I saw in Texas. The big predators living in zoos are known for doing a huge amount of pacing, and almost all of these animals were born in captivity. It's good that they were born inside zoos because it's horribly stressful for a wild animal to be captured and put in a zoo. But lions and tigers that grow up inside zoos often pace their enclosures for hours and hours.

The tiger I saw was born in captivity, but he didn't have any stereotypies at all. That's probably because his captive environment was highly stimulating. The tiger was raised by two ranchers who found him at an emu auction when he was a baby. The wife saw the tiger and said, "I'm taking him home." This was an eight-week-old male tiger cub.

They took the tiger cub home, and he lived in their house with them like a pet, becoming house-trained just like a dog. He would stand at the door to go out to go to the bathroom. The couple also owned a mature Labrador retriever who was immediately dominant over the baby tiger. After the tiger had lived with them and the Labrador for a while, they got a St. Bernard who was also dominant over the tiger. A house with two humans and two dominant dogs isn't a natural environment for a tiger cub, but it's not a deprived, barren environment, either. In the wild, tiger cubs live with their mamas and their brothers and sisters for a year while they learn how to hunt. The two dogs were the pet tiger's brothers and the ranchers were probably his parents. The tiger baby was growing up in an enriched social and physical environment.

When he reached the age of one and a half, the ranchers moved him out of the house and into a cage outside, about sixteen feet wide by fifty feet long, and he's been there ever since. They never let him roam outside the cage, but there's a little door big enough for his head to come out and they pet him and feed him. They don't have the dogs anymore, so he's pretty much alone.

That tiger has no stereotypies: no fur pulling, no paw biting, and no pacing. The only thing wrong with him is that he's a little paunchy because when he was young they overfed him, so now that he's lost weight his skin hangs down. But that's all.

The tiger has tons of cattle to look at in the pasture across the way from his enclosure, and he looks at them constantly. He gets really excited when the cattle are rotated to different pastures. If little kids visit the ranch, he also likes to look at them. He looks at small children in a really scary way—he looks at the kids the same way he looks at the cattle. That's because he wasn't raised with kids, just with grownup people and dogs. So, to him, the little person and the big person are not the same thing.

I've been doing a lot of consulting work with zoos since *Animals in Translation* came out, so I've seen a lot of big cats in captivity. This tiger looks fine to me, and if you took his cortisol levels I bet they'd be normal. (Cortisol is a stress hormone.) His current environment seems to be OK for him, but the most important thing is that he had an enriched social and physical environment when he was a cub. There is something neuroprotective—protective of the brain—about early stimulation.

To improve welfare in captive-born animals, people need to give them enriched environments both as babies and throughout adult life. It's much better to prevent stereotypies from developing in the first place, instead of trying to treat them once they've started. Once stereotypies do develop, you should try to reduce them, even in the case of scar-on-the-brain-type stereotypies. An animal like Luna may not be suffering, but the constant stereotyping itself interferes with an animal's quality of life and her nervous system is operating in a totally abnormal manner. If I had been allowed to do stereotypies all day, I would have never become a professor and I would have missed many wonderful experiences. The people who ran Luna's shelter did manage to get her stereotypies down somewhat by moving her to a different pen away from the food preparation area. The sight of food was probably making her stereotypies worse because it constantly stimulated her SEEKING system.

Everyone who is responsible for animals — farmers, ranchers, zookeepers, and pet owners — needs a set of simple, reliable guidelines for creating good mental welfare that can be applied to any animal in any situation, and the best guidelines we have are the core emotion systems in the brain. The rule is simple: Don't stimulate RAGE, FEAR, and PANIC if you can help it, and do stimulate SEEKING and also PLAY.[23] Provide environments that will keep the animal occupied and prevent the development of stereotypies.

In the rest of the book I'm going to tell you what I know about how you can do that.

2 A Dog's Life

DOGS ARE VERY DIFFERENT from a lot of other animals we work with because they are hyper-social and hypersensitive to everything we do. Dogs are so tuned in to people that they are the only animals that can follow a person's gaze or pointing finger to figure out where a piece of food is hidden. Wolves can't do it,[1] and neither can chimpanzees.[2]

Dogs are genetic wolves that evolved to live and communicate with humans. That's why dogs are so easy to train compared to other animals. Anyone can teach a dog to sit and shake hands, and most dogs do a lot of self-training as they get older. I know a dog who, every time his owner puts her shoes on to take him for a walk, runs up to her side, sits, and waits quietly for her to put on his collar. When his owner picks up the collar he bows his head. No one trained him to do any of those things. He trained himself.

The reason dogs can train themselves to perform a lot of behaviors is that our social reactions are reinforcing to dogs.[3] To train a cat, you have to give it food treats, but a dog is happy when you're happy. Over time that dog noticed that his owner acted happy when he waited quietly for his collar, so he learned to wait quietly to make her act happy.

Of course, some trainers would say that the dog trained *her* to act happy when he did his collar thing and they'd probably be right. People and dogs unconsciously train each other all the time. The natural state of life for dogs is to live with people.

Researchers have known about dogs being genetic wolves for only about ten years now.[4,5] That discovery has probably increased people's interest in the similarities between dog behavior and wolf behavior. The problem is that people have a lot of misconceptions about wolves. One of my biggest surprises doing the research for this book was reading L. David Mech's thirteen-year study of the wolves on Ellesmere Island in the Northwest Territories (now part of Nunavut) of Canada.[6] Mech's findings turn practically everything we thought we knew about wolves upside down. Since dogs are genetic wolves, that means we need to think about dogs in some new ways, too, which is what I'm going to talk about.

Dr. Mech's most important finding for people thinking about wolves and dogs: *In the wild, wolves don't live in wolf packs, and they don't have an alpha male who fights the other wolves to maintain his dominance.* Our whole image of wolf packs and alphas is completely wrong. Instead, wolves live the way people do:[7] in families made up of a mom, a dad, and their children. Sometimes an unrelated wolf can be adopted into a pack, or one of the mom's or dad's relatives is part of the pack (the "maiden aunt"), or a mom or dad who has died could be replaced by a new wolf. But mostly wolf packs are just a mom, a dad, and their pups.

The reason there is only one breeding pair isn't because the alpha male controls all the females the way a dominant wild stallion controls the females in a herd of wild horses. It's because the wolf cubs don't mate with each other or with their parents. Also, the breeding wolf pair is dominant the same way moms and dads are dominant in human families. The mom and dad never stop being the mom and dad. A fifty-year-old CEO running a major corporation is not the boss of his mom. Wolf families are the same way. The parents are always the parents, and the pups don't challenge their parents for dominance over the family.[8]

The reason everyone thought wolves live in packs led by an alpha is that most research on the social life of wolves has been done on wolves living in captivity, and wolves living in captivity are almost never natural families. They are groups of unrelated animals, put together by humans, that have to come up with some way of dealing with each other. The wolves' solution is a particular form of a dominance hierarchy, with one alpha couple who are usually the only wolves allowed to breed. This doesn't happen in the wild because in the wild nobody forces a bunch of unrelated wolves together in a pack.

Probably the other reason everyone thought wolves live in dominance hierarchies is that dominance hierarchies are extremely common in nature and in zoos and domestic settings. Most animals that naturally live in groups of adults—wild horses, for example—create dominance hierarchies, and so do domestic animals that are thrown together by their human owners.

Wolf families don't need a dominance hierarchy to keep the peace. Dr. Mech doesn't know whether the pups develop a rank order or not, and there are some researchers who think they do. But if wolf pups *do* have social rankings inside the family, they don't seem to fight each other to get the rank they want. Dr. Mech says, "Pups defer to adults and older siblings in the same automatic, peaceful way."

Dr. Mech thinks the mom probably is subordinate to the dad, although the mom's subordination wasn't obvious in the pack he observed. But even if the wolf mom is subordinate to her mate, her lower status is pretty minimal compared to the kind of subordination women have had to their husbands in the past and still have in many places. The mother wolf always greets her mate when they've been separated by adopting a subordinate posture, but her subordinate posture also causes the dad either to drop the food he's holding in his mouth or regurgitate food for her and the pups to eat. So a lot of the mom's submissive behavior is actually *food-begging* behavior, Dr. Mech says. The mom and dad work together as a team to bring down prey, and they eat their prey side by side. When the mom has really young pups, the dad is subordinate to her. And all of the wolves in the family defend their food, even down to the smallest cub in the litter. Dr. Mech says all wolves have an "ownership zone" around their mouths: "Wolves of any rank could try to steal food from another of any rank, but every wolf defended its food." That usually doesn't happen in a dominance hierarchy. One other thing people probably got wrong: the "lone wolf." Lone wolves are usually just young wolves who've left their parents and are looking for a mate.

The crazy thing about all this is that Dr. Mech wasn't the first person to say that wolves live in families, not packs. His oldest citation of a publication with this observation goes clear back to 1944, to a man named Adolph Murie, who wrote a book called *The Wolves of Mount McKinley*.[9] I think it's a really interesting question why Adolph Murie's observations didn't catch on with the public, and the captive wolf research did, especially since the wolf family idea makes so much more sense. A pack of predator animals the size of wolves is going to be hard to sustain because each animal needs so much food. Small families traveling and hunting together are much more likely to be able to kill enough prey to sustain themselves.

The huge Druid wolf pack that was put together in captivity and released to Yellowstone Park in 1996 collapsed partly because the wolves couldn't kill enough elk to feed the whole pack. At first, the pack got bigger because three females gave birth.[10] Five years after the Druid pack was released to the park, it had thirty-seven wolves. But by 2003 it was down to just eight wolves, which allowed the wolves to spread out and hunt different areas of the park.[11]

The other reason why it makes sense for wolves to travel in families, not packs, is that wolves, being predators, don't need a pack for protection the way many prey animals do. It can take a very long time for new research to filter out to the public, so I'm going to be interested to see whether Dr. Mech's article, which came out in 1999, will win out over the wolf pack image this time around.

Do Dogs Need Alphas?

Practically every dog-training book you look at tells owners that the single most important thing they need to do is establish themselves as the pack alpha. Cesar Millan's extremely popular books and TV shows are all about pack leadership.[12] But if dogs are wolves, and *wolves* don't have pack leaders, why do *dogs* need a pack leader?

I think it will take a while for researchers and trainers to figure out all the ways that a set of wrong ideas about wolves probably created a corresponding set of wrong ideas about dogs. Many behaviorists and ethologists who specialize in dogs have protested against the "alpha male/dominance" image of dogs. But they haven't given a lot of thought to wolves living in families and what that means for dogs.

Dogs evolved to live with humans, but what does that mean? Did dogs evolve to live with human *families*? And if they did, does that mean dogs living with human families need a mom and a dad, not an alpha? Or are dogs living with human families more like a forced wolf pack than a family, in which case somebody has to be the alpha?

I'll start with the case for owners being the alpha.

Cesar Millan and His Dogs

Cesar Millan's huge collection of thirty to forty dogs living together at his Dog Psychology Center in Los Angeles is an extreme version of the artificial wolf packs humans have created with captive wolves in research centers and shelters. Since captive wolves form dominance hierarchies when they're forced to live together, it would be natural for Cesar's dogs to form a dominance hierarchy, too, with an alpha dog, a beta dog, and maybe even a third-ranking dog, too. Instead of letting that happen, Cesar made himself the alpha.

Cesar says all dog owners have to be the pack leader. If dogs living with human beings are *psychologically* like wolves living with unrelated adults, then he's probably right. The question is: Are adult dogs living with human families like adult wolves living with unrelated wolves, or are dogs living with human families like wolf cubs living with their parents?

Cesar bases his ideas about dog packs on the groups of working dogs that lived on the farms in Mexico when he was a child. The dogs stayed together in groups of five to seven animals. They didn't live inside the house and they weren't pets, but they weren't feral, either. They were working dogs that helped herd cows, guarded property, and protected the humans.[13]

Those dogs don't sound like they were living in dog families. Cesar says they usually came from different breeds, and most of the packs had dominance fights, which a family of wolves wouldn't have.[14] That brings up another interesting aspect of Cesar's childhood experience with dog packs that made a big impression on him. The dogs in his grandfather's pack didn't fight for dominance:

> Some of the other ranchers seemed to have dogs with fairly tight pack structures, where one dog was pack leader and the others were followers. Those families liked to watch when their dogs got into battles over dominance — when one dog beat another down . . . I had also witnessed dominance displays in the feral dog packs that ran wild in the fields near our house.[15] But . . . the dogs on our farm didn't seem to have a discernible pack leader among them. I realize now that this was because my grandfather never let any dog take the leadership role away from him — or from the rest of us humans, for that matter.[16]

Cesar might be right about why his grandfather's dogs didn't fight. I think Cesar is *definitely* right about the big group of abandoned and aggressive dogs at his Dog Psychology Center. Cesar's work with those dogs shows that it's possible and probably desirable in a lot of cases for humans to become the pack leader of an unrelated group of dogs. When the human is the pack leader, the dogs don't fight.

But I can't tell what was going on with the dogs on the grandfather's farm. Those dogs might have seen the grandfather as pack leader. On the other hand, they could have seen the grandfather as their human parent. The reason they didn't fight could have been that they were treating each other like wolf cub siblings, and wolf cub siblings don't fight.

I say that because the dogs on the grandfather's farm sound a lot like the dogs I grew up with. There were four dogs in my neighborhood who lived with four different families, and since there were no fences around any of the houses, the dogs were usually hanging out or playing in the same physical space. Three of them were intact males (no one neutered males in the 1950s) and one was a spayed female: two were golden retrievers (our dog, Andy, and another retriever named Lightning) and two were Labs (Hunter and Tucky). The dogs slept and were fed in their respective homes but ran loose during the day.

Usually we kids would be riding our bikes out in the street, which was a quiet circular drive, and all four dogs would be running around us. They never got in fights, and they never chewed up the house because they weren't inside the house during the day. They were outside, where they had things to do. Barking was seldom a problem except for one weird experience we had with our dog, who sometimes stayed out late and barked under the window of one particular house. The people who lived there didn't have a dog, so that couldn't have been what was bothering Andy. The only thing unusual about those neighbors was that my parents had gotten into several arguments with them.

Sometimes the dogs left and we didn't have any idea where they were. There was a giant estate across the street from my house that probably had a hundred acres of woods, so I imagine they spent a lot of time there. But I don't know if they went there together or separately because I never saw them go or come back. They'd just be around sometimes, and not be around other times.

The dogs were almost always together when they were around us kids, but I don't remember them being like a pack with an alpha, though I do remember dominance displays. Whenever our golden retriever, Andy, would see Lightning, the other golden, he would instantly lie down on his back for a few seconds. Sometimes he'd go over to Lightning's yard and voluntarily lie down in front of him. I never saw either of the Labs do that. (There's a very good reason why I didn't, which I'll get to in the next section.)

Nobody really knows what a natural state for a dog is, partly because the human lifestyle keeps changing. But I think the dogs in my neighborhood were probably living in the most natural state there is for a dog. The town dogs in the book *Merle's Door* by Ted Kerasote sound like the dogs in my neighborhood. Kerasote's town doesn't have any fences, and the dogs all seem to get along very well. None of them barks when another dog walks past the house, which is hard to imagine when you live in a town where yards are fenced in. Instead, the dogs smell each other's butts and wag their tails.

The reason I think the most natural existence for a dog is a fence-free, mostly outdoor life with a human owner is that this is probably the way dogs lived with people a hundred thousand years ago when wolves first evolved into dogs. That was long before agriculture was invented (just ten thousand years ago), and humans lived as nomadic hunter-gatherers. Hunter-gatherers probably had a home base to come back to after the hunt, but the home base moved around a lot because humans had to follow food sources.

The first dogs might have had some type of "owner" or human family, but they wouldn't have lived in homes with fences. (There's no way to know whether early dogs were ever restrained, or whether they were completely free to come and go.)[17] I think those early dogs probably acted like the dogs I grew up with. They went hunting with the humans, or went off roaming on their own, then came back to the home base. At some point they probably stopped living in dog families and started associating as small groups of unrelated adults. But they weren't a pack, and they didn't have an alpha. They were coworkers, playmates, and probably sometimes traveling companions. Their main attachment was to their human families, not to each other.

My theory is that Cesar Millan might be wrong about the farm dogs he grew up with, but he's right about the dogs at his Dog Psychology Center. His grandfather's dogs probably weren't a true pack. The dogs at the Dog Psychology Center, though, are living in one of the most unnatural situations any dog can live in: thirty to forty unrelated animals and different breeds all stuck together inside one big enclosure. Those dogs *have* to form a pack to keep from ripping each other apart, and Cesar has done a brilliant job of making himself the pack leader instead of letting the dogs fight it out for the top spot. More than half of Cesar's dogs are homeless animals, most of them with aggression problems, and they are all living together peacefully.[18] Cesar Millan has made an unnatural situation work well for his rescue dogs.

Dogs Need Parents, Not Pack Leaders

What dogs probably need isn't a substitute *pack leader* but a substitute *parent.* I say that because genetically dogs are juvenile wolves, and young wolves live with their parents and siblings.

During evolution dogs went through a process called pedomorphosis, which means that dog puppies stop developing earlier than wolf cubs do. It's a kind of arrested development. That's why dogs — especially purebred dogs — look less "wolfy" than real wolves. Baby animals have "baby faces" the same way human babies do. Newborn wolf puppies have little snub noses and floppy ears just like newborn dog puppies, but the wolf puppy grows up to have a long, pointy nose and tall, pointy ears. Most dogs grow up to have shorter, snubbier noses than wolves, and a lot of dogs have floppy ears like a puppy's ears, too. Purebreds are especially young-looking. A friend of mine says they have "toy faces."

Thanks to some really interesting research done in England, we know that dog facial features and dog behavior generally go together. Dr. Deborah Goodwin and her colleagues found that the more wolfy a breed looks, the more grown wolf behaviors it has. To study the connection between wolfy looks and wolf behaviors, she chose the fifteen most important aggressive and submissive behaviors wolves use to communicate with each other during a conflict, and then observed ten dog breeds to see which breeds expressed which behaviors. Aggressive behaviors included things like growling, teeth baring, "standing over" (one dog puts its head over the other dog's body), and "standing erect" (the dog stands as tall as it can, with its back arched and its hackles up). Submissive behaviors were things like muzzle licks, looking away (the submissive dog averts its eyes and very slowly turns its head away), crouching, and the *passive submit*, where the dog lies on its back and exposes its anogenital area.

Dr. Goodwin found that Siberian huskies, which of the ten breeds look the most like wolves, had all fifteen behaviors, whereas Cavalier King Charles spaniels, which look nothing like wolves, had only two.[19] The correlation between looking like a wolf and acting like a wolf was pretty strong across all ten breeds, with some interesting exceptions. Three of the four gun dogs — cocker spaniels, Labrador retrievers, and golden retrievers — had somewhat more wolfy behaviors than their appearance predicted, and two of the sheepdogs — German shepherds and Shetland sheepdogs — had somewhat fewer wolfy behaviors than their pointy noses and ears predicted. The German shepherd and Shetland sheepdog are probably the exceptions that prove the rule because their facial features were deliberately bred into them starting with sheepherding stock. The German shepherd was intentionally bred to look as much like a wolf as possible. Dr. Goodwin says that may mean that once a breed has lost a behavior you can't bring the behavior back just by changing its appearance. So although looks and behavior go together genetically, they can also be separated genetically. She thinks the reason the gun dogs kept as many wolfy behaviors as they did might be because hunting dogs need "a fuller range of ancestral behaviour" to do their job.[20]

Even with the exceptions, the overall order supported her hypothesis:

1. Cavalier King Charles spaniel: 2 wolf behaviors out of 15
2. Norfolk terrier: 3 of 15
3. French bulldog: 4 of 15
4. Shetland sheepdog: 4 of 15
5. Cocker spaniel: 6 of 15
6. Munsterlander: 7 of 15
7. Labrador retriever: 9 of 15
8. German shepherd: 11 of 15
9. Golden retriever: 12 of 15
10. Siberian husky: 15 of 15

It would be interesting to do Dr. Goodwin's study using mixed-breed dogs. Mutts revert to a somewhat wolfy body form fairly quickly, but do they also get some of the wolfy behaviors back? No one knows.

When you think about dogs being wolves that haven't finished growing up, people who treat their dogs as if they're children might have the right idea after all — although that doesn't necessarily make them good "dog parents." Also, people who buy lap dogs and treat them like babies are probably right for at least some of the highly neotenized toy breeds that have retained puppylike behavior. Dr. Goodwin says that a King Charles spaniel never matures mentally beyond the stage of a puppy. It even looks very much like a puppy after it is full grown. I saw an adult Cavalier at the airport once while I was waiting to catch my flight. Everybody was coming up to pet this darling, puppylike dog.

If dogs need parents, does this mean people should throw out their guidebooks on how important it is to establish themselves as the alpha dog?

I think it depends on the book. Some of the guides have probably been right for the wrong reason. Dog owners do need to be the leader, but not because a dog will become the alpha if they don't. Dog owners need to be the leader the same way parents do. Good parents set limits and teach their kids how to behave nicely, and that's exactly what dogs need, too. Dogs have to learn good manners and their owners have to teach them. When dogs don't have good human parents, they get crazy and out of control and take over the house the same way an undisciplined, spoiled child gets crazy and out of control and takes over the house. It probably doesn't matter whether you think of yourself as the alpha or as the mom or dad so long as you raise your dog right. And because a dog never does grow up mentally, you have to keep on being a good parent and setting limits even after your dog is grown up physically.

One way or another, the human has to be in charge. Whether

you think of yourself as mom, dad, or pack leader probably doesn't matter as long as you're handling your dog right.

How Many Dogs Are Too Many?

I think Dr. Mech's research gives us a possible biological reason why it's a very good idea not to own any more than two dogs unless you know what you're doing. The dog behaviorist Patricia McConnell says, "People who have more than one dog are in a special club . . . you know that two dogs are more than twice as much work as one, and that three dogs are as much work as you expected seven to be."[21]

How well a group of dogs will get along depends on both the personalities of the dogs and the personality of the owner. Some types of dogs will be easier to handle in groups than others.

Several dogs living inside one house with one owner is a forced pack, and forced dog packs may cause trouble. Dr. Dorit Urd Feddersen-Petersen has done studies where she grouped together a number of dogs from the same breed and let them roam free inside a large open-air enclosure. She wanted to see how they would get along under the "seminatural" conditions of their "ancestral species," the wolf.

They didn't do very well:

> We found that some dog breeds are unable to cooperate (in a very basic manner: just doing things together) and compete in groups, reflected in difficulties in establishing and maintaining a rank order (e.g. poodles). The interactions in these dog groups are not functional, and the members have difficulties coping with challenges from the environment. It is striking that ways . . . of conflict solving (to appease, animate or inhibit the opponent), a common practice in wolves, do not exist in groups of several dog breeds . . . Within many groups of dogs, trivial conflicts often escalate into damaging fights.[22]

These were purebred dogs put together in groups with other purebred dogs. They couldn't manage conflict because they had lost a lot of the submissive behaviors wolves use to keep conflict from escalating to a bloody battle.

The worst breeds were toy poodles, West Highland white terriers, Jack Russell terriers, and some genetic lines of Labrador retrievers. The breeds that got along best were malamutes, huskies, Samoyeds, German shepherds, American Staffordshire terriers, golden retrievers, and Fila Brasileiros. The breeds that did get along pretty well had more social tolerance of each other than the toy poodles and terriers, made more nonaggressive approaches to each other, and did more *allogrooming* (one animal grooming another).

Dr. Feddersen-Petersen's study probably fits with Dr. Goodwin's research because Dr. Feddersen-Petersen also seems to have found a difference between the more wolfy breeds and the less wolfy ones. The wolfy breeds such as malamutes, huskies, and Samoyeds did much better living in forced packs than breeds that are (probably) less wolfy, like the toy poodle and Jack Russell terrier.

Her results make sense when you think about *which* wolf behaviors the less wolflike breeds have lost. What they've mostly lost, out of the fifteen most important aggressive and submissive behaviors wolves use to communicate during a conflict, are the submissive behaviors. The less wolfy the dog, the fewer submissive behaviors it has. The King Charles spaniels in Dr. Goodwin's study didn't have a single one of the six submissive behaviors wolves use during conflicts, but the Siberian huskies had all of them. The six breeds with seven or fewer wolf signals had either one or no submissive behaviors, whereas the four breeds that were closer to wolves had several submissive behaviors.[23] Labs and German shepherds showed three of the six submissive behaviors, goldens showed four, and huskies showed all six. The fact that some of the Labs in Dr. Feddersen-Petersen's study couldn't live together peacefully is exactly what you'd expect from a breed that's lost half of its submissive behaviors. Wolves can use more submissive behaviors to stay out of a fight.

It might seem surprising that wolves have many more submissive behaviors than dogs, given how often you hear that wolfdogs (wolf-dog hybrids) are more dangerous than regular dogs. Young wolves develop aggressive behaviors first and then develop the submissive behaviors later on, but dogs stop developing before all of the submissive behaviors adult wolves use to get along with each other have come in. So why aren't dogs more dangerous than wolfdogs?

When it comes to aggressive versus submissive behaviors, a wolf cub's development is probably similar to a human child's development. A young cub, like a young child, can get away with more aggression than adults can because a wolf cub or a child can't do that much damage. The aggressive behaviors come in first, so young wolves (or young children) have some way to defend themselves if they have to. The submissive behaviors come in second, so an older, bigger wolf or human has ways to *stay out of fights* with other juveniles or adults. We definitely see that in normal humans. A normal two-year-old child may hit his mom; a normal twenty-year-old would never do such a thing — at least, no normal human who's been well brought up.

For a dominance hierarchy to work, the other animals have to do two things: first, *agree* to be subordinate, and second, *signal* their agreement through submissive behaviors. If a breed has lost the submissive behaviors that wolves need to form stable hierarchies, then the dogs probably won't be able to avoid fights unless they have also been bred to be nonaggressive. Dr. Feddersen-Petersen also found that groups of dogs fight less when humans "participate more frequently in group life."[24] A frequently absent owner and a forced pack of adult purebred neotenized dogs is a recipe for disaster.

I've always noticed that packs of dogs are the most dangerous.

We don't have very good statistics on dogs and dog aggression in general, but at least one study of fatal dog bites, published in 1983, reported that dog pack attacks on humans made up 18 percent of all fatal dog attacks, but only 1 percent of nonfatal dog bites.[25] These two studies give us part of an explanation for why that would be. Dr. Feddersen-Petersen observed group aggression in some of her purebred dog packs, where "many group members joined a collective attack on a threatened animal."[26] The only known occasion on which wolves have made a vicious group attack on a lone wolf happened in the Druid pack in Yellowstone Park.[27]

Very likely, when you put a group of unrelated adult purebreds together in a pack, you're playing with fire. They don't have all the submissive behaviors and signals wolves do, *and* they're living in the one situation where even wolves might launch a group attack on another wolf. Now that these two studies have been published, the advice I would give anyone who wants multiple purebred dogs would be to have a maximum of two unless the dogs are part of a family such as mother and daughter.

Using the Blue-Ribbon Emotions to Create a Good Environment for Your Dog — Are Two Dogs Better than One?

Dogs are so emotional and expressive that it's fairly easy to see how Jaak Panksepp's blue-ribbon emotions apply. Since I've just been talking about how many dogs a person should adopt or buy, I'll start with PANIC, which is the social attachment system inside the brain. PANIC is what makes baby animals cry when their mothers leave; when you run a weak current through the PANIC system in the brain, the animal makes *separation calls.* Animals and people cannot be happy if their PANIC system is activated. For animals to be happy, their social needs have to be met.

So, what are a dog's social needs?

We know dogs need people because they evolved to live with people — but do dogs need other dogs, too? It's obvious dogs like other dogs and enjoy being with their doggie friends. In the past, dogs have always been around other dogs; it's natural for a dog to spend some time with other dogs, whether they live together or not. I worry about the fenced-in lives of dogs today. Family dogs aren't free to come and go the way the dogs I grew up with were and the way Merle is in *Merle's Door.* That's the whole point of *Merle's Door.* Merle is a free dog and he has the behaviors of a free dog.[28]

This is a huge change in the lives of dogs from just twenty or thirty years ago: dogs aren't free anymore. I don't think anyone knows what the effect has been. I believe that if you did an epidemiological study of dog-directed aggression, you would find there's more of it today than there was when I grew up. I can't tell whether there's definitely more human-directed aggression, although dog bites did increase by 36 percent between 1986 and 1994.[29] An increase in the number of dogs does not explain this statistic because dog numbers increased only 2 percent during this time period. Dog-directed aggression and human-directed aggression are at least somewhat separate traits genetically, so dog-to-dog aggression could have risen without dog-to-person aggression also going up.[30] My question is: Are we seeing an unintended consequence of leash laws? By passing laws to make life safer for dogs, did we make it more dangerous for people?

Leash laws put Cesar Millan and all the other experts who stress pack leadership in a different light. They're talking about today's dogs, not about the dogs that lived twenty or thirty years ago. Although dogs evolved to live with people (and people probably evolved to live with dogs), dogs aren't people, and a dog whose life is totally controlled by human owners might be psychologically like a wolf living in a forced pack. If leash laws and fences have made dogs feel more like members of a forced pack than a family, maybe today's dogs do need an alpha.

That brings up the question of how much time a dog can be left alone and still have good emotional welfare. It seems likely to me that one dog can live happily never or almost never seeing or interacting with other dogs if its owner is around all the time. But dogs are too social to be happy staying alone for hours on end. Veterinarians and trainers hear hundreds of stories about dogs going crazy inside their owner's house when the owner is away at work. I've seen videos of dogs chewing up doors and window blinds trying to get out of a house where they're locked up alone. No dog does that while its owners are home. Furniture chewing happens because of separation anxiety. The only time that Andy chewed anything was when he was a puppy. He chewed up some shoes and my sister's stuffed animals.

This is another bad effect of today's leash laws and fenced yards. It's almost as if dogs have become captive animals instead of companion animals, and the house or fenced yard has become like a really fancy zoo enclosure. So, when you buy a dog today, you have to think about how to make up for the fact that he's not going to live the life that comes naturally to him.

I think that if you have to leave the house all day long to go to a job, either you shouldn't get a dog, or maybe you should get two dogs, preferably two dogs that know each other. Another alternative would be to choose a dog that has lower attachment needs and less PANIC emotion activation. He may be more likely to sleep during the day.

I hear about an awful lot of dogs that are suffering from loneliness and boredom. Dogs may have even less ability to tolerate being left alone for a long period of time than a wolf might because dogs are juvenile wolves, and juvenile wolves stay with their parents until they're around two years old. Some wolf juveniles have been known to stay with their parents until they were almost three.

The little lap dogs, such as the King Charles spaniel, may be even more needy because they're mentally younger than the big gun dogs. I mentioned that the King Charles stops developing dominance and submission behaviors at a wolf age of just twenty days. That's too young for a wolf or a dog to be spending a lot of time on its own, and some behaviorists who specialize in dogs are starting to confirm this. *New Scientist* magazine quoted one, Donna Brander, as saying, "These juvenile types have a problem with separation anxiety . . . They have never gone into a more mature mode of behaviour."[31]

So I strongly suggest that if you're going to be away a lot, or can't pay an hour's worth of attention to your dog every day, you should consider getting two dogs. A lot of people think having two dogs is more fun than having just one anyway, and watching dogs play is a blast. But if you're determined to get just one dog, and you have to be out of the house all day long, find a good doggie daycare place to take your dog to while you're gone.

Forced Pack?

Does doggie daycare replicate a forced pack similar to the ones that caused problems for the wolves? I think a forced pack may create problems for some dogs that tend to be more aggressive, but others get along really well. Many dogs love going to the dog park to romp and play. There are three factors that will determine whether or not a forced pack will be a problem: genetic tendency toward aggression, lack of hard-wired submissive behavior, and lack of early socialization with other dogs. Many Labs are so nonaggressive that all they want to do is play, so the lack of submissive behavior has no effect. The occasional Lab who is aggressive can be a real problem because some genetic lines have dog social problems. I have observed that dogs that are reared to adulthood in isolation are often vicious with other dogs.

Dogs are also very specific in how they think. The behavior of a dog on a leash toward other dogs may be very different from his behavior off the leash. When he is on the leash he may be motivated to protect you, and when he is off the leash, it's play time. Research on the brain shows that it creates "file folders" like a computer. Certain situations are in the "protect my owner" file (FEAR) and others are in the fun (SEEKING) file.

Dog Socialization — Not Just for Puppies

The other big issue with the PANIC system is a dog's ability to get along with people and other dogs. Patricia McConnell makes an excellent point in her book on dog emotions: Socialization is not the same thing as enrichment. You need both.[32] Puppies need to be socialized to other dogs, cats, children, and adults between the ages of five and thirteen weeks. That's the sensitive period, and if you wait until puppies are older, they'll never be as well socialized as they could have been.

There are a lot of good books on how to socialize your puppy, so I won't go into it here. You don't need a lot of advice anyway; it's mostly common sense. The most important thing is to expose a puppy to children, adults, and dogs that live outside the household as well as to family members.

The one thing I do want to add, which the books don't seem to cover, is that teenage dogs must have a second round of socialization. Dr. Karen Overall, a research veterinarian who specializes in the new field of behavioral medicine,[33] says dogs become sexually mature between six and nine months, but don't mature socially until eighteen to thirty-six months. A lot of dog trainers think this in-between period is similar to the teen years in humans, when young dogs develop rambunctious energy and test their limits.[34] Based on what we know from other species, which I'll talk about in a later chapter, I think it's a good idea to have your dog spend time with some adult dog role models during those months, especially if your dog is a male. All the dogs from my childhood learned social skills from other adult dogs.

When Andy was a teenager, he freely interacted with adult males. Lightning and Hunter were his teachers.

RAGE — Dogs Flying off the Handle

Jaak Panksepp says that the RAGE system probably originates in the experience of being physically restrained.[35] The feeling of frustration that comes from *mental* restraints, like a dog not being able to get out of a fenced yard when it sees a squirrel to chase, is a mild form of anger.

All animal owners and caretakers should create environments that are as low-frustration as possible, but puppies need some deliberate exposure to frustrating situations so they can develop frustration tolerance. Puppies have to learn how to control their emotions and their actions just like children do. Patricia McConnell, who works with problem dogs, says she's had hundreds of owners come to her office saying their dogs are "out of control," which means that the dogs go straight from frustration to rage to aggression.[36] These dogs have no tolerance for frustration.

All dogs that are neurologically normal can learn how to control their anger, but they do have to *learn*. Learning emotional restraint is easy for some dogs, who figure it out from their everyday environments. But if you've got a dog who seems emotionally reactive, you should give him some positive experiences with mild frustration. Frustration is part of life, so reactive dogs and people have to learn to keep their frustration from escalating to rage. They also have to develop impulse control so that even if they do feel intensely angry, they don't do anything about it. A safe dog is a dog that doesn't fly off the handle when the neighbor boy sticks his hands in its face and screams in its ear. There are plenty of dogs like that; it's one of the reasons people love dogs so much.

A puppy's first, and maybe most important, lesson in frustration tolerance happens in the litter. Dr. McConnell says she's seen what she thinks is a disproportionate number of *singleton* puppies (puppies born in single births) with serious behavior problems. Singleton puppies can grow up to be abnormally aggressive.[37]

Dr. McConnell tells a very interesting story about a singleton puppy she raised. She was extremely worried about how he would turn out, so she tried to duplicate the conditions of having littermates. One thing that happens to all puppies is that they have to climb all over each other when they nurse, and they get knocked off the nipple a lot by their brothers and sisters. So Dr. McConnell used a little stuffed toy to push the singleton puppy off the nipple for a couple of seconds when he nursed. Unfortunately, this approach didn't seem to work as a substitute for having real siblings, because when he was just five weeks old the puppy growled at Dr. McConnell one day when she touched him. Dr. McConnell says, "A five-week-old puppy growling at a person is like a five-year-old child stabbing his mother with scissors. On purpose."[38]

She was scared to death about how that dog would turn out, so after that she spent a lot of time training the puppy to like being touched. Every time she touched him briefly, she'd give him a treat, which is classic positive reinforcement. That did work, and eventually the puppy started asking to be touched. When he had learned to behave like a normal dog, Dr. McConnell gave him to a woman who didn't have children, and everything worked out great.

The Dog Who Gave Himself a Time-out

If you watch a normal, well-behaved family dog, you'll probably see lots of examples of frustration tolerance and impulse control, especially in well-behaved dogs that aren't naturally cheerful. I know a dog like that. He's a brown and black mutt who has Rottweiler, pit bull, and hound ancestors. He is a very somber dog. He almost never has the *open-mouth play face* you see on a happy, relaxed dog, and his eyes usually look soulful or even a little sad. His owner remembers him having the same look on his face at the shelter where she picked him out. He's the opposite of a happy-go-lucky dog.

As a puppy, this dog showed a lot of bad signs. He growled at the children in the family, and one time he growled at a six-foot-five plumber. The plumber thought it was cute that such a little dog wanted to take him on, but his owners knew it wasn't. Worse yet, a growling puppy was especially bad for this family because one of the kids was autistic and highly unpredictable.

The owners read a couple of books about dominance and dominance aggression, and they figured they had to make sure the dog knew he was subordinate to every member of the family, including the autistic boy. So they didn't tolerate any growling under any circumstances. The few times the dog growled at one of the kids, the mom angrily pushed him away from wherever the growling had taken place. She didn't hit him, but she yelled and pushed him roughly.

This is a situation where the idea of *dominance* could hurt more than it helps because it's a misdiagnosis of the problem. The problem with this dog wasn't that he was *dominant,* although he was acting like the kind of dog the books call dominant. The problem was that he was *emotionally reactive.* I don't think the mom needed to do lots of dominance displays with the dog; she just needed to reward him when he was being calm, and administer a firm but not loud consequence when he growled. Fortunately, dogs are so tuned in to people that with most of them you have a lot of leeway for mistakes. Dogs want to please you, and this dog got the message that growling at humans was strictly against the law.

One interesting thing about this family's approach, though, was that they ended up giving the dog a lot of training in frustration tolerance without realizing that's what they were doing. They did this partly because they were worried about their autistic son, and partly because the puppy's social behavior seemed "off" to them. They'd gotten the puppy in the first place because their old dog had died and their seven-year-old child wanted to get a puppy. But this puppy wasn't interested in either of the kids and wasn't even that social with the grownups in the household. In the evening, instead of hanging out with the family, he would go lie down in his crate in another bedroom down the hall.

The mom didn't think this was good. She also felt that the family dog has a job to do, which is to play with the kids or at least not freak out when the kids want to play with *him*. So in the evening she would go and get him out of his crate and bring him up onto the bed where she and her husband and one or two of the kids would sit to watch TV. She wouldn't make him stay for hours, but she did bring him in. She would put him next to her, which he liked; then, when it seemed like he was getting tired, she would take him back to his crate or just let him go on his own.

She also did all the usual things a lot of dog-training books tell you to do, such as taking food or toys away from him, handling his mouth and looking at his teeth, moving him when he was asleep, and so on. She did all of these things in a friendly way, and he didn't have any problem.

She thought she was doing dominance training, which in a way she was. (I'll talk about dominance when I get to FEAR.) But what she was really doing was using positive reinforcement to train a very uncheerful little puppy to deal with frustration. That puppy learned frustration tolerance and impulse control so well that he figured out how to give himself a time-out if he needed one. One day, for instance, she'd taken the dog out for a walk with the next-door neighbor and her two dogs. By then the family had gotten a second dog, so there were four dogs on the walk altogether. Previously the mixed breed had gotten into two serious fights with one of the neighbor dogs, but after the second fight both dogs had learned how to go on walks together without any problem.

That day, all of the dogs except for the mixed breed started playing together in the snow. They were really worked up, and the mutt was getting more and more excited watching them. He was barking his head off.

He was also trying to keep his distance. Every time the three dogs got close to him he'd back up, and if they got really close he'd race up to the top of a seven-foot-tall pile of snow that had been left by the snowplow. Then he'd stand up there watching the other dogs and barking from on high. After a while he'd come back down and bark at the other dogs from closer range, but the minute they started getting too close again he'd put himself back up on top of his snow mountain.

Both of the neighbors interpreted his behavior the same way — he was preventing any of the other dogs from ramming into him. The second fight with the neighbor dog had happened when that dog — a golden retriever — had come bounding out of some thick bushes without knowing the mixed-breed dog was there, and the two had collided. They were both so surprised that their impulse control didn't have time to kick in, and they instantly started to fight. The owner thinks her dog knew he might lose his temper if he got slammed into, so he was deliberately taking himself too far away from the action for that to happen. He was giving himself a time-out.

That dog is six years old now and has been a very good member of the family who gets along great with the kids and is always happy when company comes over. He's even pretty good with other dogs once he's been around them for a few minutes. This is a dog a lot of people would have said had a very good chance of not working out.

I think the dog's early environment made the difference. Some families, when they heard the dog growling at their child, would have trained the child to give the dog a lot of space. But because this family knew they wouldn't be able to train their autistic child well enough to rely on him to stay out of the dog's way, they had to train the dog instead. He grew up to be a dog who could monitor his emotions and stay out of trouble.

How to Train Your Dog to Tolerate Frustration

Dr. McConnell says the two best ways to train puppies to tolerate frustration are to teach them the "stay" and the "wait" commands. It's especially good to teach dogs to "wait" for a couple of seconds before letting them go out the door. Dr. McConnell says even a microsecond is OK when your dog is young and rambunctious. A lot of dogs get into an uproar racing out the door for their walk, so that's a good time to teach your dog to moderate his emotions and mind his manners. The "door wait" is a training exercise a lot of the dominance books have misinterpreted. A trainer who stresses dominance and pack leadership will tell you always to go in or out a door before your dog. But that's not necessary. The important thing about the door wait is waiting. Once your dog has waited quietly for a second, it doesn't matter who goes out the door first, you or your dog. The dog has been reinforced for showing impulse control and emotional restraint, and that's all that matters.

It's also a good idea to train puppies not to mind when you take food away. Wolves never give up food if they can help it. Dr. Mech once watched a wolf dad who hadn't eaten for quite a long time spend an hour trying to steal a bone from his mate, who had just eaten a big meal of meat off the bone. The mom had no reason to hold on to that bone, but she snapped at him and didn't let him take it. That bone was hers, and she wasn't letting it go. If you train a puppy to let you take food away, you've built in very high frustration tolerance.

Labs and Other Family Dogs

I mentioned that some dogs learn frustration naturally in the course of everyday life with their human families. However, you shouldn't assume that you don't have to think about training for frustration tolerance just because you've bought a family-friendly breed. I'm thinking about Labrador retrievers in particular.[39] There are two kinds of Labs. There is the great big heavy-boned, heavyset Lab that is content to lie around all day. I call this personality type the "wheelchair temperament," because these dogs are so calm they make excellent service dogs for people with handicaps.

If you have a Labrador with the wheelchair temperament, you probably don't have to worry about training for frustration tolerance. But there's another kind of Lab, too: the hyper kind. These dogs can be great with children because they're innately cheerful and incredibly energetic. A friend of mine who has a yellow Lab named Molly told me her thirteen-year-old son said, "Molly has three expressions: happy, a little bit less happy, and a little bit more happy." That's a good description of a hyper Lab's natural temperament. You can tell that a dog is happy when its mouth is in a relaxed, partially open to fully opened position.

Hyper Labs can be a lot of fun and they love to fetch or play Frisbee. If they get lots of fun exercise they are less hyper. But they don't have great emotional restraint or impulse control. That's pretty much the definition of "hyper": lack of impulse control. Anyone who's ever owned a hyperactive Lab knows what I'm talking about. Molly is so hyper that no one in the family could pet her until she was almost five years old, and even now it's not easy. She gets so revved up over the wonderfulness of being petted that she can't sit still.

I think that might be why Dr. Feddersen-Petersen found that only *some* Labs — not all — showed severe aggressive behavior when she put them in forced dog packs. The aggressive Labs may not have had the emotional restraint to keep their conflicts from escalating into RAGE and violent aggression. So even though Labrador retrievers are one of the safest breeds there is, it's a good idea to train any dog to "wait" and "stay." It's probably a good guess that the hyper-type Lab is more likely to lose control and bite, but I don't know this and I don't think anyone else does, either. The RAGE system is universal, so all dogs should learn emotional restraint.

Knowing FEAR When You See It: Is Dominance Aggression an Anxiety Disorder?

Dog trainers have talked about *dominance aggression* versus *fear aggression* for years, but most owners find the distinction extremely confusing when they try to use it to deal with a dog that's showing aggression.[40] Fear aggression is easy to understand; fear aggression happens when a frightened dog feels cornered and lashes out. The signs are easy to recognize: when a dog with fear aggression sees a person or situation that scares him, he'll back away while growling and barking, and once he can't back up any farther he may bite. Fear biting is what separates the fear-aggressive dog from a normal dog that's frightened. A normal dog puts his tail between his legs and tries to run away when he's scared, but he doesn't bite.

The idea that there are two completely different kinds of aggression gets really confusing when you're trying to deal with dog-to-dog — or *interdog* — aggression. What's the emotion motivating one dog to attack another dog? Is it fear? Is he trying to protect his owner or his territory? And if he's trying to protect his owner or his territory, what emotion system does that come from?

All actions are driven by emotion, so to understand dominance aggression we have to figure out what emotion is driving it. The best explanation I've seen is the section on dominance aggression in Dr. Karen Overall's textbook on the clinical treatment of behavior disorders in cats and dogs. Dr. Overall says dominance aggression comes from an underlying *anxiety disorder* in the dogs. They're not *afraid* so much as they are *anxious.*

There's a lot of interesting research being done on anxiety disorders in human psychiatry that is going to apply to dogs, too, so dog behaviorists' thinking about anxiety in dogs will probably be changing in the next few years. For now, we can say that anxiety is related to fear, though it's probably not just a milder form of fear because there are some biological differences between fear and anxiety.[41] But to the degree that anxiety is related to fear, dominance aggression is related to fear aggression.

The difference between classic fear aggression and dominance aggression is that a fear-aggressive dog just wants to get away, but a dominance-aggressive dog is anxious about his *control* over resources (food, toys, sleeping spot) or behaviors (running off leash, getting up on the bed, etc.). The reason that he doesn't want to run away isn't that he's fearless and dominant, but that running away won't solve his problem. For example, if a dog gets anxious when you touch his food bowl, he's not going to feel better if he runs away and lets you take away his food bowl. The only thing that will make him feel better is for you to stop touching his food bowl. That's why he growls: to get you to go away and leave his food bowl alone. Yes, he's protecting his things, but it's not because he's dominant and wants control over you. He's *anxious.* You might say he has *anxiety aggression.*

From the way Dr. Overall describes dominance-aggressive dogs, they sound like humans with obsessive-compulsive personalities. This is the kind of person you would call "controlling" or "micromanaging." I think she's right about this because dominance-aggressive dogs are very tense animals. They're the opposite of relaxed and happy.

The other evidence for the idea that fear aggression and dominance aggression are both related to fear (or anxiety) is that they often respond to the same antianxiety treatments. These treatments include medications, such as amitriptylene,[42] an antidepressant, along with some behavioral treatments.

Deep Pressure Is Calming

I'm interested in a new physical treatment, called an *anxiety wrap,* that was developed by a dog trainer named Susan Sharpe after reading about the squeeze machine I made when I was an adolescent to soothe my anxiety. I've written about the squeeze machine a lot before. I came up with the idea after seeing cattle being put into a squeeze chute that held them still so they could get their shots. When I saw how calm the cattle got from the pressure on their bodies, I built my own squeeze machine and it calmed my anxiety the same way. Using that same idea, Susan Sharpe created a kind of T-shirt for dogs that applies snug pressure across the dog's body. She says it can help with all kinds of problem behaviors, including phobias, fear, and aggression.

Another dog behavior specialist, Nancy Williams, has had good success wrapping the midsection of the dog with the wide elastic bandages that are used to wrap a horse's legs.[43] My friend tried the pressure treatment on a hyper Wheatland terrier who could not tolerate the hustle and bustle of a Thanksgiving family gathering. She wrapped her arms around the dog's midsection and squeezed him. Afterward he was calmer and more attentive.

I've also seen some amazing research on three aggressive Great Danes, two with dog-to-dog aggression and one who was aggressive to strange people, who were put in something called full-body restraint and then exposed to strange dogs or humans. *Full-body restraint* is done by putting an animal inside a box with its head sticking out of a hole in the front, and filling the box with oats so that the animal's entire body up to the neck is encased in grain and it can't move. Full-body restraint makes animals incredibly calm, and all three dogs became totally calm even when looking straight at another dog or human. These were dogs with severe aggression problems. The next time they saw the grain box they happily went inside, and they transferred their new ability to be calm instead of aggressive to their homes. One of the dogs was still dramatically improved years later.[44]

With any kind of pressure treatment, you have to be careful not to leave it on too long. The maximum calming effect wears off in about twenty minutes, so, for longer treatments, it often works best to apply the treatment for twenty to thirty minutes, take it off for thirty minutes, and then reapply.

Until we learn more, I think dog owners should assume that a lot of unexplained aggression in dogs has a basis in fear or anxiety, and they should take steps to relieve that painful emotion. Instead of focusing on dominance, which isn't an emotion, they should focus on two things:

- identifying and treating the aggressive dog's fear and/or anxiety
- training the aggressive dog for emotional restraint and good manners

A Pushy Dog Isn't a Bad Dog

The other important thing to realize is that *dominance* isn't the same thing as *dominance aggression.* All dominance means is that when two animals (or two people) want the same thing, the dominant animal is the one that gets it. The dominant animal usually doesn't use aggression to get what it wants, and in a true social hierarchy the dominant animal is almost always less aggressive than lower-down animals.[45]

This means that dominant behaviors don't automatically lead to dominance aggression, and in a lot of cases aren't connected to dominance aggression at all. You have to go by the individual dog's personality. A happy-go-lucky pushy dog isn't a threat. A somber, growling pushy dog is. It's the emotion motivating the behavior that counts.

So much of the advice on "establishing dominance" over your dog is wrong because it's not tied to the dog's emotions. An example is dog-training books telling you never to let a dog up on the furniture or on the bed. Dogs on furniture are up too high, supposedly. Another example: books telling people never to play tug of war with their dogs because playing tug of war encourages the dogs to think it's OK for them to challenge you. That's completely wrong. In *Animals in Translation* I wrote about the study of golden retrievers playing tug of war. The experimenter played tug of war with fourteen dogs, letting one group win almost all the games while the other group lost most of the games. All of the dogs were more obedient after playing tug of war, regardless of whether they won or lost. Beating a person at tug of war didn't make the dogs more dominant.[46] I wouldn't be surprised if you saw the same thing in a study of dogs getting up on furniture and beds.

The Dog Who Learned to Heel

Even though a lot of dog-training books focus on dominance and being the pack leader, you can get a huge amount of good advice from them if you remember one thing: You are the grownup and you need to stay calm if you're going to teach your dog to stay calm. All good trainers, behaviorists, and ethologists will tell you how important it is to maintain control of your own emotions when you're dealing with dogs. Cesar Millan calls it being "calm assertive," and Patricia McConnell says, "Dogs seem to love people who are quiet, cool, and collected and prefer sitting beside them over sitting beside others."[47]

Once you know that your goal is to teach your dog impulse control and emotional restraint, you can pick and choose which pieces of advice to follow from different trainers. My friend with the mixed-breed dog solved a pretty serious interdog aggression problem by accidentally applying a core principle from Cesar Millan: Teach your dog to heel. Cesar says Americans let their dogs walk in front of them, pulling on the leash, and that makes the dogs dominant. That's why they get in fights with other dogs on walks.

My friend applied this principle to her dog accidentally because she hadn't read Cesar Millan or seen his shows and she hadn't succeeded in training her dog to heel. Her dog wasn't good with other dogs. If a stranger dog they met on a walk approached her dog and tried to sniff or greet him, her dog would attack. My friend coped with this by taking walks where they weren't too likely to meet other dogs. Of course they did still cross paths with other owners and their dogs once in a while, and if the other dog approached her dog there would be trouble. She tried different things to try to fix the problem, such as having her dog sit when another dog was approaching, but nothing worked. The dog just seemed to get worse, to the point where he was going crazy just *seeing* another dog.

But then, quite a while after she'd given up on changing the dog's behavior, she realized that he wasn't going as crazy as he had been. She had no idea why, but she started to notice that instead of getting worse and worse, her dog was getting better and better. He could even let a strange dog sniff him sometimes.

She finally figured out what had happened when she read an interview with Cesar Millan in which he said that dogs that don't heel develop behavior problems because they're too dominant. All of a sudden, my friend realized what the difference was: After my friend started using an electronic leash so her dog could run loose without taking off, her dog had taught himself to heel. Cesar's prediction was right; as soon as her dog started to walk beside her or behind her, he started to lose his aggression toward other dogs.

A lot of people object to electronic leashes, or *gun-dog collars*, because people use them to train dogs by giving them lots of shocks. I agree with them; you shouldn't use shock to train a dog. But the gun-dog collars have a beep function, too, which is what my friend used. Her house has a wireless fence and the beep the gun-dog collar makes is the same one the wireless fence makes. The dog had already learned after being shocked just once on the electronic fence that the beep was a warning. As soon as she put the gun-dog collar on her dog and he heard the beep, he decided that the safest place to be was right next to or just behind her, so he just naturally started to heel.

After my friend read the interview with Cesar, she interpreted her dog's improved behavior as the result of her becoming dominant. I think what really happened was that the dog realized he felt much more secure walking behind his owner instead of out front where he could see strange dogs walking straight toward him. Face-to-face isn't a normal way for dogs to approach each other. If you watch Cesar Millan's shows, when he introduces two dogs to each other, he always has them walk side by side before letting them get into each other's faces, and Dr. McConnell gives the same advice for humans meeting a potentially dangerous dog. Don't go toward a dangerous dog face-to-face, and never make eye contact. Primates like face-to-face introductions; dogs don't.

Purebreds versus Mutts

My friend thinks part of the problem her mixed-breed dog was having with the dogs they met on walks might be that most of those dogs were purebreds. Many times she would notice a purebred dog bounding up to her mutt, not seeming to notice that her dog's body was stiff and his hackles were raised. (Sometimes the owners seemed pretty oblivious, too.)

Dr. Goodwin's study showing that purebred dogs have lost a lot of the natural submissive behaviors of wolves might mean that my friend's dog really *was* being provoked by at least some of the purebred dogs—lots of Labs, quite a few small terrier breeds, a few golden retrievers and poodles—they met on walks. The purpose of submission signals from one wolf to another is to keep the peace. The message is, "I don't want to fight." But my friend's dog was constantly being approached face-to-face by dogs showing very few polite, submissive behaviors. It was obvious these dogs didn't want to fight her dog because they always acted surprised and frightened when her dog attacked. But they seemed to have lost most of their ancestors' ways to signal their peaceful intentions.

I wonder whether some purebred dogs haven't lost not only the ability to do some wolf behaviors but also the ability to read some standard wolf behaviors. Is it possible some purebred dogs don't know what raised hackles mean?

Another question is whether mutts might be more anxious than purebreds in general. Mutts may be closer to wolves genetically because as soon as purebred dogs mate outside their breed, they start to develop physical features more similar to those of wolves: pointier noses, darker fur if the breed is yellow or white, and so on. If mutts are also more similar to wolves behaviorally, they're going to be more anxious than a lot of the purebreds because wolves are extremely shy. Wolves are so shy that researchers have been able to crawl inside a wolf den and handle the cubs while the wolf mom and dad hide from the humans.[48] Dogs are much less shy than wolves, which is almost the definition of a dog versus a wolf. Wild animals that have been domesticated are less shy around humans than their wild counterparts, or they'd still be wild. So by definition mutts can't be as shy as wolves. But mutts may be shier overall than purebreds.

There's some research to support this. Greg Acland, at Cornell, crossbred shy Siberian huskies with two generations of beagles and got two very shy male dogs in the second litter. He interpreted his results to mean that shyness appears to be transmitted as a dominant trait.[49] If shyness is a dominant trait, then mutts should start to regain that trait as they crossbreed.

Dogs Are Individuals

For years dog owners have been told that "any dog can bite." Owners are supposed to be super-vigilant about establishing dominance over their dogs, because the dogs might become dangerous if they don't. But telling people that any dog can bite is very misleading because it lumps all dogs together. The chances of a completely normal, well-socialized dog that hasn't been traumatized as a puppy biting a person are tiny.

Dogs have different personalities, different temperaments, and different biologies. I'm sure we'll find out that aggressive dogs have a different neurological makeup from nonaggressive dogs. In her textbook, Dr. Overall writes about the results of an unpublished study she and her colleagues did at the Veterinary Hospital of the University of Pennsylvania (VHUP). They tested the urine of 210 aggressive dogs and compared it to the urine of 84 nonaggressive dogs. Eighty-eight percent of the aggressive dogs had metabolic abnormalities compared to only 23 percent of the nonaggressive dogs.[50] The most common abnormalities were very high levels of glutamine, taurine, and alanine. Glutamine and taurine are *excitatory neurotransmitters*,[51] and high levels of glutamine have been associated with high levels of aggression running through human families.

Criminal Dogs

You have to know your dog, and you have to know your dog as an individual, not just as a member of a particular breed. Pit bulls are a good example of how important it is to take each dog as an individual. Pit bulls were originally bred to fight other dogs, but *never* to attack humans. Any pit bull that attacked humans was put down and its genes were taken out of the gene pool. Today there are illegal breeders and gang lords who are deliberately breeding the nice-to-people characteristic of pit bulls out by crossing people-aggressive dogs with dog-aggressive pit bulls. I saw one videotape of eight-week-old puppies that were already fighting viciously. That's totally abnormal. These people are deliberately breeding criminal dogs. An animal shelter lady told me even the rescue group people can't handle these dogs and are bringing them back to the shelters. I also learned about a litter of puppies sired by a vicious pit bull. The young puppies were adopted out to different families and the adult dogs were all returned for biting.

Some communities hear stories like this and conclude that pit bulls should be outlawed. But outlawing a particular breed is never the answer because it's so easy to breed dogs for aggression. If you make pit bulls illegal, unethical breeders will just breed other dogs for aggression, and it won't take them long to come up with an animal that's genetically built to be nasty. This has already started to happen. Akitas and chows are being crossed with other aggressive breeds to create killer dogs. Some pit bulls are dangerous due to breeding for extreme prey drive. I watched a pit bull at an animal shelter get a behavior test for adoptability. He passed the food-guarding test with flying colors, and he had no reaction when a fake hand was thrust into his food bowl. His reaction during the prey-drive test, however, was scary. When a life-size mechanical cat was presented, he locked onto it like a fighter jet's radar. This dog failed the test and was not eligible for adoption. All you can do with this kind of information is use it as a check on the individual dog you're interested in buying or adopting to make sure he hasn't been affected by bad breeders.

How to Spot an "Easy" Dog

Before I talk about easy dogs, I want to make sure everyone knows that any *normal* dog can be a great pet, including fearful, shy, and anxious dogs. Sometimes — not always — these dogs take more work, but they are great companions.

If you want a dog with an easygoing temperament, you can give any dog a quick personality test before you buy or adopt it. A lot of different personality tests have been developed. One of the easiest for most people to use would be what Patricia McConnell does to test a puppy's temperament. After she spends some time getting to know a puppy, here's how she tests his emotional makeup:

> . . . my favorite exercise with a young pup is to *gently* roll her over onto her back and then lightly restrain her with a hand on her chest . . . I'll let my hand rest lightly on her until she begins to try to get up, and then I'll use just enough pressure to keep her from doing so . . . Pay attention to what the pup does when you let him go. I've seen puppies who have rolled over, nailed me with one of those "looks could kill" expressions, and refused to come near me again. That's not the dog I want, because I love dogs (and people) who don't take life too seriously. Neither do I want a dog who is so stressed by the procedure that his eyes round in terror . . . My favorite dog is the one who takes the whole thing as a silly game.[52]

This approach tests a puppy for excessive fear *and* for excessive anger. It tests for fear because grown dogs and wolves never roll each other over unless they're in a really bad fight. Even though Dr. McConnell does it very gently, it's still potentially a scary thing to do to a dog. It tests for anger because anger comes from the experience of being restrained. And it tests for an easygoing nature because any puppy that is confident and calm enough to think that being rolled over and lightly held down is all a big game is going to be a very good-natured dog.

End of Life Issues

It is often difficult to make the right decision about care for pets who are nearing the end of life. How do you determine when euthanizing a pet is the right thing to do? When is it right to perform a painful, invasive treatment, and when is it right to choose a simpler treatment that will reduce pain and improve the quality of the animal's remaining life?

When you are faced with these questions, you need time to think carefully and have discussions with friends and family. If you are informed at the veterinarian's office that your pet has cancer or some other serious disease, the options for treatments vary widely. I strongly recommend taking your time to make a decision. In a nonemergency situation, bring your pet home and take several days to decide what to do. You need time to get over the shock so you can make a good decision. Discussing all the possible options and getting a second opinion from another veterinarian may also be a good idea. You must remember that animals are not able to understand that a painful treatment could prolong their life. They just feel the fear and pain. You need to look at the quality of the animal's life.

Would a complicated, invasive treatment improve the quality of your pet's life? There are also financial realities that require good judgment. Most important, you must make sure that your choice does not prolong suffering.

What Dogs Need: PLAY and SEEKING

Dog fears and aggression can be hard to figure out sometimes, but dog joy isn't. Just about anyone who's lived with a dog knows what dogs like. In terms of the core emotions, dogs need:

- social contact so their PANIC system doesn't get activated
- games and play with their owners to activate the SEEKING system
- interesting things to do — especially long walks — that arouse their SEEKING system

We've already talked about a dog's social needs. Dogs should not be left alone all day long cooped up in a house or an apartment, and they certainly shouldn't be left inside a crate for hours on end. If the house is going to be empty during the day, you should either buy another dog or find a good doggie daycare or both. Another good arrangement would be to leave your dog with a neighbor who stays home during the day. Patricia McConnell says dogs need social companionship almost as much as they need food and water.[53]

New scientific research published after the hardcover edition of this book was printed clearly shows that dogs have been selected to be more socially aware of human signals than wolves. The journal *Science* published an article titled "Going to the Dogs," which covered the rapidly expanding field of canine cognition. This research reinforces McConnell's view that dogs need social companionship. Dogs that live in animal shelters need at least forty-five minutes of play and exercise time with a person every day. My graduate student Christa Coppola conducted a study in an animal shelter. Her study indicated that when a dog had been played with for forty-five minutes its cortisol (stress) hormone was lower the next day. Unfortunately, the cortisol levels returned to sky-high levels if the play period was not repeated. Dogs in animal shelters need volunteers to play and socially interact with them every day.[54]

Dogs especially need the PLAY system to be stimulated because they never grow up all the way. All juvenile animals play more than adults. So, if you can afford it, you should buy plenty of toys for your dog, and you should rotate the toys the same way people rotate toys for their kids. Old toys are boring; new toys are fun. That's the rule. You should also play with your dog every day. If your dog likes to chase balls, that's great; if he likes to play tug of war, that's another good game.

Dogs probably have high SEEKING needs because they are descended from wolves, and wolves are nomadic animals that get a lot of mental stimulation during the day and make a lot of decisions. There is nothing dogs like more than a long walk, unless it's getting loose outside the yard and taking off for a day on their own or with a doggie friend.

A friend of mine once ran into her dog when he and the neighbor dog were on the loose. She was riding bikes with her eleven-year-old son and all of a sudden, out of nowhere, there were the two dogs, running along beside them. The dogs greeted them with huge, relaxed, open-mouth dog grins, ran alongside them for a little way, then took off up a hill, running as fast as they could. They looked like they were having the time of their lives.

Since dogs don't have many opportunities to roam free these days, it's up to you to give your dog enough mental stimulation to keep his mind busy. Dogs have to have a daily walk, and I think it's a good idea to take those walks someplace where your dogs can be off leash, if possible. They'll get a lot more exercise and do a lot more investigating and exploring that way, which is what they need to activate the SEEKING system.[55]

Patricia McConnell says dogs need at least an hour a day of attention from their owners, and that's just the average. Working breeds need huge amounts of exercise. Your one hour a day could be a one-hour walk, or, if you don't have the energy for that, you could break it down into three parts: a half-hour walk, fifteen minutes of play, and fifteen minutes learning new tricks.

Teaching dogs new tricks is especially important for exercising their minds and activating the SEEKING emotion. Some high-energy dogs love to do agility training and you may want to join a local dog training group, or set up some equipment such as jumps and tunnels in your yard.

Dogs need people, play, and lots of opportunities to explore and learn, and they can't provide these things for themselves.

That's your job.

3 Cats

THE BIG DIFFERENCE between cats and dogs is that cats aren't hyper-social. You can't use social approval to train a cat, and cats don't train themselves by picking up on their owners' reactions the way dogs sometimes do. Dogs serve people, but people serve cats.

On the other hand, cats are *not* solitary, self-sufficient loners the way a lot of people think. Cats have social needs. Unfortunately, we animal behaviorists and ethologists don't know as much about cats and their emotions as we do about other domestic animals. But a lot of what we do know hasn't gotten out to the public.

One of the most important things to realize about cats is that they haven't really been domesticated, at least not nearly to the degree dogs have.[1] Wolves started evolving into dogs a hundred thousand years ago. No one knows for sure yet when wild cats started to evolve into domestic cats. The oldest cat remains found in the grave of a human are 9,500 years old. That was after agriculture was invented in some parts of the world, so humans were past the hunter-gatherer stage. They were living in towns and villages when they started to associate with cats. The most popular theory of how cats became domestic animals is that they joined human settlements to prey on rats and mice. Basically, they made themselves into pets.

This theory may not be right, though. James Serpell, associate professor of humane ethics and animal welfare and director of the Center for the Interaction of Animals and Society at the University of Pennsylvania School of Veterinary Medicine, thinks hunter-gatherers may have had cats, too. Archaeologists found a cat jaw in one of the first human settlements on Cyprus, when agriculture was just being invented. If the very earliest villages had cats, then hunter-gatherers probably had cats, too. Dr. Serpell says that the few hunter-gatherers who are still alive today like to capture young wild animals and care for them, which is evidence that "animal keeping" has always been practiced by humans. Contemporary hunter gatherers take really good care of these animals, don't eat them, and mourn them when they die.

If Dr. Serpell is right, cats and people could go back a very long way. But even if cats and people have lived together for thousands and thousands of years, cats probably haven't been changed that much by their association with people because cats and humans had a *mutualistic* relationship instead of the more *symbiotic* relationship humans and dogs had during domestication.[2] Early humans needed their dogs to guard their camps and help them hunt, and early dogs needed their humans for food and shelter. They depended on each other. With people and cats, it was more a relationship of convenience. Cats killed mice and rats, and humans provided lots of mice and rats to kill since mice and rats lived in human settlements. The two species didn't need each other so much as profit by being around each other.

The result was that today a housecat is a lot closer to a wild cat than a dog is to a wolf. To understand why, you need to know how domestication works. The first wolves who started to live with humans would have been less fearful of humans than other wolves. Those less fearful wolves would have been fed by humans, which would have given them a reproductive advantage over their brother and sister wolves who still had to find all their own food. That created a selection pressure for tameness.

For example, if the slower antelopes are the ones that get eaten by the lion, the faster antelopes will be more successful at breeding and will leave behind more offspring because they live longer. After a few generations, if the slow antelopes keep getting eaten at higher rates than the fast antelopes, antelopes will become faster as a species. Selection pressure is the mechanism through which animals evolve.

Humans who kept wolves would have gotten a reproductive advantage, too, because they had wolves guarding them while they slept and helping them hunt. After a few generations, the less fearful wolves would be on the path to becoming domestic dogs, and the humans would probably have been on their way to becoming a species that likes to live with dogs.

Cats would have had less selection pressure to lose their fear of humans than wolves did, partly because the African wild cat, which is the ancestor of the domestic cat, has a lot less fear of humans than wolves do. There have been reports for 150 years of African wild cats living on the edges of villages and of the people in the villages capturing the cats as kittens, taming them, and using them to catch rats and mice. Europeans also wrote about taming wild cats and keeping them to catch rats.[3] The African cat didn't have to change very much to be willing to live as a housecat.

Cats did change in some ways. All domestic animals have smaller brains than their wild ancestor animals, and domestic cats also have smaller brains than wild cats although we don't know very much about the differences at this point.[4] Domestic animals also undergo *pedomorphosis*, or *neoteny*, and adult housecats have three juvenile behaviors: meowing, purring, and kneading their paws. Adult wild cats living in zoos don't do these things with humans, although they do purr with other cats.[5] Other than that, though, cats are less neotenized than other domestic animals, and they can easily go feral and survive. That's not true for dogs. If you put the family poodle out in the countryside, his chances of surviving are low unless he finds another family to live with. But abandoned cats that are used to living outside may do just fine. They're healthier living with people because they get veterinary care, but they don't need to find another family to survive. They are still adapted to living in the wild and taking care of themselves. However, cats that have lived indoors all their lives may die because they have never learned to hunt or fend for themselves.

There was a second set of selection pressures on wolves and early dogs that cats didn't have, which was that humans started preferring wolves that were *especially* good at guarding the settlement or going on hunting parties with them. That would have started dogs down the path of becoming specialized for work. Cats weren't big enough to guard the humans from big predators, so there wouldn't have been any selection pressures turning them into guard cats, and there wasn't any reason for humans to choose the best mice and rat killers because all cats naturally kill mice and rats and any other small prey they can catch.

Even without knowing anything about the history of the domestic cat's evolution, you can tell from the cat's appearance and behavior that cats haven't changed as much as the other domestic animals have. In appearance, domesticated animals are much more varied than their wild counterparts. Dogs have huge variability, ranging all the way from tiny toy breeds to Alaskan malamutes. Cats don't come in anything like this kind of variety. They've evolved some different coat colors, but most people looking at photographs of wild African cats wouldn't be able to tell them apart from a standard tabby cat.

There are at least three reasons why domesticated animals have more variability in appearance, which probably apply more weakly to cats than to the other domestic animals. First, humans usually protect domesticated animals from predators, so selection pressures for protective coloring go down because genes for coat colors that make the animal easier to see don't get culled out of the gene pool. Second, the genes for appearance are usually connected to behavior in some way, so as different behaviors develop in a domesticating animal, appearance changes, too. I'll talk more about the relationship between a cat's coat coloring and its behavior later on.

The third reason why there's so much variability in domesticated species is that humans deliberately breed domesticated species to be more variable. That's the second stage of domestication. At first domestication happens naturally when some of the animals living with humans (the good guard dogs and the good hunting dogs) reproduce more successfully than others (the guard dogs and hunting dogs that got killed). Then, once humans developed selective breeding techniques, they began to breed differences into domestic animals on purpose. Dogs and horses were bred to do particular jobs; cows, goats, sheep, and chickens were bred for production traits such as how fast they grew, how much meat and fat they had, and so on. People have been selectively breeding cats only for the past 150 years or so, and purebred cats are being bred for appearance traits instead of work or production traits.

There's also a big behavioral difference between cats and the other domestic animals, which is that you can't train a cat using punishment and negative reinforcement. That makes cats more like wild animals than dogs or horses or cows. Karen Pryor has a really good description of how wild animals react to punishment or force:

> Anyone who has ever kept a wild undomesticated animal for a pet knows that they are more difficult to train. It is extraordinarily difficult, for example, to teach a wolf to walk on a leash, even if you have raised it from puppy hood and it is quite tame. If you pull, it pulls back automatically, and if you are too insistent and pull too hard, the wolf, no matter how calm and sociable it usually is, panics and tries

to escape. Put a tame pet otter on a leash, and either you go where the otter wants to go, or it fights the leash with all its might.

Dolphins are the same. Push a dolphin and it pushes back. Try to herd dolphins from one tank to another with nets; if they feel crowded, bold individuals will charge the net and timid ones will sink to the tank bottom in helpless fear.[6]

The only way to train a wild animal is to use positive reinforcement. Positive reinforcement means rewarding the animal for doing the things you're training it to do. Cats are the same way, which is probably why people have always thought cats weren't trainable. Traditional animal training relies on punishment and negative reinforcement. The punishments and negative reinforcements might be extremely mild, but they're still negative. For example, putting a collar on a puppy and pulling on the leash to get him to walk with you creates a slightly unpleasant pressure on the puppy's neck. The puppy learns to walk forward to *get away from the pressure.* That's what negative reinforcement is: an animal or person is reinforced by having something negative (pressure on the neck) *stop.* (I'll talk more about the difference between punishment and negative reinforcement in the next chapter.)

You can't train a cat to walk on a leash by putting a collar on it and dragging on the leash. But you *can* train a cat to walk on a leash by using positive reinforcement. I'll explain how later on.

What it all comes down to is that a cat in a person's house isn't all that different from a cat on the Serengeti Plains. Nicholas Dodman, the veterinary researcher who wrote *The Cat Who Cried for Help,* says, "A cat is in some ways like a miniature tiger in your living room."[7]

Cats Are Hard to Read

I started out this chapter by saying cats are more social than most people realize. One of the reasons people haven't picked up on cat sociability is that domestic cats aren't totally domesticated the way dogs and horses are. They can go their own way. Quite a few cats *do* go.[8] A friend of mine told me that her favorite cat when she was growing up, a big, striped tomcat named King, moved to the neighbor's house down the road. They'd see him every once in a while when he came back to visit, but otherwise he stayed with the neighbors. My friend grew up on a farm where all the cats were barn cats, so her parents figured King must have gotten promoted to housecat with his new family. He had a better offer and he took it. A dog would never do that.

The other reason people see cats as being more solitary than they are is that cats have less in common with people as a species than either dogs *or wolves* do, regardless of domestication. Wolves live in families with a similar structure to the human family, and wolves' communication is probably a lot more like human communication than cat communication is. Wolves communicate heavily through vision, and their faces are very expressive. Thanks to domestication, dogs also make lots of different sounds people interpret pretty well. For instance, the reason dogs "talk" — or bark — is because they evolved to live with people, and people communicate through sounds. Wolves are mostly silent. Adult wolves almost never bark.

Dogs also read humans well. There's a lot of research on that.

Cats are completely different, and I think their differences make them difficult for people to read. The hardest thing for people is that cats don't have expressive faces. Humans naturally look at an animal's face to see what it's thinking because people are primates, and primates use their faces to communicate. Some anthropologists think facial expressions are more important than words, even. Looking at an animal's face works just fine with a dog, although you need to look at its posture and its tail, too. But cats don't signal with their faces very much, and they have lots more bodily signals than either dogs or wolves do.[9] So when people look into their cats' faces, they're looking at the wrong place.

Another interesting thing about cat faces: Cats don't have eyebrows the way people and a lot of dogs do. Eyebrows probably evolved to highlight facial expressions, and a lot of dark-furred dogs have little round light spots right over their eyes that may have evolved for the same reason.

I think the cat's inexpressive face is one of the reasons some people think autistic kids are like cats. There's even a book called *All Cats Have Asperger Syndrome.* Cats seem autistic because they don't come across as being sociable or eager to please like dogs, and also because their faces are kind of blank. Autistic people often have somewhat blank faces, too.

Cats probably don't read people's faces very well, either. In his book about cats, *The Cat Who Cried for Help,* Nick Dodman has a story about a cat who was badly scared one day when he saw a man beat his dog right outside the sliding glass window that the cat was sitting next to. The man was tall, thin, and had a beard. Later that day the cat attacked his owner, who was also tall, thin, and had a beard, so it's possible he mistook his owner for the other man.[10] A dog wouldn't have made that mistake.[11]

The final reason people have trouble understanding cats is that humans probably can't even perceive a huge number of the signals cats make. Cats use smells to communicate with each other, and humans have a terrible sense of smell compared to cats. Researchers say cats may have as good a sense of smell as dogs, who can smell things at a threshold a thousand times lower than people can. Of course, everyone can smell it when a cat has been spraying the furniture, but there are probably all kinds of subtler smell signals inside those big, horrible smells that other cats can pick up on but we can't.

Cats definitely communicate with other cats by leaving odors humans can't smell. They have at least two ways to do this. One: Cats have glands on their paws that leave a deposit whenever

they scratch an object. Two: Cats also have facial g
leave deposits when they rub against things. Dogs m
ing on stuff, but that's all. They don't rub up again
knead people with their paws to leave their scent.

Given the fact that cats do so much "talking" in smells, it isn't really surprising that the number-one behavior problem cat owners go to vets for help with is elimination disorders.

That brings me to the core emotion systems in the brain.

The Blue-Ribbon Emotions in Cats: The FEAR System

Fear can be a big problem for cats, which is probably why we have the expression *scaredy-cat.* A lot of housecats are so afraid of strangers that they hide anytime someone comes to visit. I wonder whether this is because cats are genetically closer to their wild counterparts. A wild animal is afraid of people. That's one of the things that makes it wild.[12] It's possible cats naturally have more fear of humans than dogs do because they're not as fully domesticated.

It's easy to tell when a cat is afraid. Fearful cats avoid the thing they're afraid of however they can. A housecat runs and hides. At the zoo, when the big cats are afraid, they refuse to go out to the exhibit area. A cat who is only feeling fear doesn't puff out his hair,[13] arch his back, and spit. That's an angry cat. The FEAR system can activate the RAGE system in all animals and people, so a frightened cat can become enraged. But as long as he's *only* feeling fear, he doesn't puff up his hair or attack.

Three different studies have found that cats can be divided into two personality categories: bold and shy.[14] The researchers who did the three studies called the bold cats *confident, easy-going, sociable,* and *trusting;* they described the fearful cats as *timid, nervous, shy,* and *unfriendly.*

Boldness and friendliness go together. Bold cats are faster to approach a novel object than shy cats, and they are friendlier to people. So, if you want a *naturally* friendly cat (later I'll talk about things you can do to make a shy cat friendlier), you could consider getting a cat from a breed that's known for being friendly. The Siamese breed is a good bet. In one survey of people who owned Siamese cats, Persian cats, or nonpedigree cats, both the Siamese and the Persian were ranked as being friendlier than the domestic cats. The authors of the survey also observed the owners with their cats and found the same differences the owners reported.[15, 16]

If you're going to get a domestic cat, there are three things you can do to obtain a cat with a friendly, bold temperament:

1. *Get a kitten and make sure lots of people gently handle it when it's tiny.* The sensitive period for socialization is the second week of a kitten's life to the seventh week, and the more people who handle the kitten during this time, the better. This is very important because it's so easy for cats to go feral. My friend Mark's cat once had kittens underneath the house where he couldn't find them. By the time he finally discovered where they were, the kittens were way over two months old and were super-wild. He couldn't handle them.

   I've seen the opposite happen, too. I went to a livestock auction on a day they weren't having a sale, and I found a mama cat and her kittens living underneath the auctioneer's stand. Those kittens had been handled by everyone at the auctions and they were the friendliest, cutest little kittens. Kittens need a lot of friendly human handling during the sensitive period.

   Handling won't turn a genetically shy cat into a completely friendly one, though. At least two different studies show that shyness is inherited, and although you can reduce it some, you can't totally reverse it.[17]

2. *If you adopt a kitten from a shelter, pick one that is friendly.* I've gone to a number of animal shelters and visited the cats. If I put my hand in the cage, some kitties come right up to me and rub me while other kitties huddle in the back. The kitty that comes right up to you is the kitty you want.

   A lot of shelters have a room where you can get to know the animal, so you can watch to see how a cat acts when you take him into the room. If the cat lets you hold him, that's a good sign. You could also take a cat toy in to see whether the cat plays with it. There's a wonderful little cat toy I really like: a little feather duster attached to a flexible eighteen-inch wand. Cats just love it.
3. *Adopt a black cat.* Sarah Hartwell, a shelter worker in England, calls black cats "laid-back blacks" and tortoiseshell cats "naughty torties." That description is supported by a handful of studies showing a relationship between fur color and behavior. Black cats especially are friendlier than other cats, are better able to deal with crowding and urban life, and have greater *aggregative tendencies,* which means they're more inclined to live in groups of cats. Black cats are more social overall, whether it's with other cats or with humans.

Black fur on cats has an interesting history. Black fur coloring is caused by recessive genes. For anyone who's forgotten high school biology, a recessive gene has to be inherited from both parents if the offspring is going to have the trait. Blue eyes are caused by a recessive gene. The black fur gene has evolved four different times in four different cat species. Obviously, black fur — or the traits associated with black fur — gives cats an important survival advantage. Stephen J. O'Brien with the feline genome project thinks the mutation that causes black fur might make cats resistant to viruses in the HIV family.

Other studies have found that orange male cats are more aggressive than black male cats. That's logical because orange cats are shier than black cats, and you would expect fearful cats to have more fear aggression. I've noticed that neutered orange males and females can be very affectionate. Some orange cats will rub on you all day. However, orange cats startle and scare easily.

The black cat's calmer nature might explain why there are so many black cats around even though the gene for black fur is recessive. Black cats that live in cities are more successful at mating than orange cats because orange cats spend too much time fighting the other tomcats, whereas the black cats patiently wait their turn to mate.

It's different out in the country, where there aren't as many cats around. Orange cats have an advantage in rural areas because they can monopolize the local females without having to fight a lot of other male cats. Country cats observe a form of *polygyny,* which means having more than one wife at the same time, whereas city cats are mostly promiscuous.[18] Different genetics work well in different situations.

Fur color isn't a guarantee that a cat will have one kind of personality or another, so adopting a black kitten doesn't mean you've definitely got a cat that likes to live in groups, including groups of humans. When you choose a kitten, you have to go by the individual personality of the kitten regardless of color.

Preventing Fear at the Vet's Office

The most frightening place for cats, even bold ones, is the vet's office. To keep a cat calm for medical exams and treatments, you handle it the same way good stock people handle a cow, using the principles of restraint I developed for cattle:

- No sudden jerky motions — use calm, steady movement.
- No slippery metal tops on the examining table. I tell

people to bring a bathmat with a rubber backing from home to put on the table. Slipping causes panic in all animals.

- Stroke your cat firmly as a way of applying deep pressure. *Do not* use pats or light tickle touches.

A student once came up to me after I finished my lecture and said he worked as a veterinary aide. Everyone at his clinic called him the "kitty whisperer," he told me, because he was the only person who could calm the cats down when they came in for their appointments. The secret of his success was my principles of restraint for cattle, which he'd read about in *Animals in Translation.* He tried applying them to cats, and they worked!

Kittens and cats should be trained to feel that a cat carrier is a safe place. Food treats can be fed in the carrier and your cat should be gradually taught to tolerate being locked in the carrier for a longer and longer time. This prevents the car ride to the vet from being a yowling, frightening experience. Your cat should become fully accustomed to the cat carrier *before* he has to go to the vet or go on a car trip.

When I was a child, we did everything wrong when traveling to our summer house with Bee Lee. Both my mother and I thought that Bee Lee would travel easily like our dog, who just hopped into the car so he could be with us, but Bee Lee had never been trained to stay in a carrier, and he ripped apart the flimsy cardboard box I had put him in. On another trip, he clawed Mother's head as she was driving down the freeway. Mother had him riding loose because he yowled in the carrier. She mistakenly thought he would be calmer when loose. A bad experience like this can create a permanent fear memory and make it very difficult to train a cat to travel calmly in a carrier. Cats that have had a scary travel experience will probably need tranquilizers before their next trip.

Elimination Disorders

Most marking behavior in cats comes from anxiety, which is related to the FEAR system. Dr. Dodman says that "when cats are anxious about something they become insecure and develop a strong need to redefine their territory."[19]

Male cats are usually the problem markers, but female cats can also develop elimination disorders due to inappropriate marking. Dr. Dodman says you can tell what a cat is anxious about by noticing the areas it's marking. He's had a few clients whose cats marked *them*. One lady's cat urinated right in her face while she was lying in bed.[20] The cat was also peeing on the lady's clothes, backpack, and purse. The cat was upset because the lady's boyfriend had moved in with her. Dr. Dodman told her that the cat was marking her and things that belonged to her "to let the world know that you are hers."[21] The cat completely changed on BuSpar, an antianxiety medication. She stopped marking her owner's things, and she started liking the new boyfriend and playing with him.

Cats can get *interstitial cystitis*, which is similar to the kind of urinary tract infections people get. However, when Dr. Overall studied cats with cystitis at the Veterinary Hospital of the University of Pennsylvania, she found that only 15 percent of them also had behavior elimination disorders. The rest were using their litter boxes just fine.[22] The antidepressant Elavil has been a standard treatment for cystitis. Nick Dodman says this might mean that interstitial cystitis is related to anxiety.[23]

Elimination disorders come in at least three different varieties:

- spraying furniture and rugs
- defecating outside the litter box
- defecating *and* urinating outside the litter box

Each category has a set of subcategories, too, such as *substrate aversion* (the cat develops an aversion to the kind of kitty litter its owner buys) and *location aversion* (the cat develops an aversion to where the litter box is). Forty to 75 percent of all cats brought to the vet for problem behaviors are spraying, urinating, and/or defecating in the wrong place.[24]

A cat can decide it doesn't like its litter box for a lot of different reasons, which I'll talk about in the next section. A lot of these problems are easy to fix by doing the following:

- Change the substrate. (Some cats prefer sandlike kitty litter and others hate strong deodorant smells.)
- Change the location of the litter box. (Your cat may think the box is too exposed or too far away from the rest of the house. As an example, a cat probably wouldn't be happy having its litter box way up in the attic or in a far corner of the basement next to the noisy furnace.)
- Change the flooring under the litter box. (Cats don't like the plastic mats people put under the box to protect the floor, possibly because they move too easily. All animals intensely dislike slippery or unsure footing. Any unstable flooring will frighten an animal.)
- Change the litter more frequently.
- If you have more than one cat, make sure you have enough litter boxes for all of them (preferably one box per cat).
- If you have more than one cat, put each cat's litter box in a different room.
- Change the type of litter box. (Some cats like boxes with hoods and some don't.)

Up until recently people had no idea how complex a behavior an elimination disorder is. Cat owners tended to see spraying and peeing or pooping in the wrong place as a nuisance (a lot more than a nuisance in severe cases), somewhat like the nuisance of having an untrained puppy in the house. But it's really an emotional problem.

An elimination disorder can mean different things at different times. A cat that poops by the door, for instance, may have seen a stranger cat outside. That turns on the FEAR system and the cat poops to mark its territory and lower its fear by marking out a safety zone. Marking territory, for cats (and dogs, too), is probably like putting up a backyard fence for humans. An ordinary fence won't keep *really* bad guys out and everyone knows it. A fence is a social signal that tells law-abiding people this space belongs to you.

Cat Obsessions and Compulsions

Cats have high levels of OCD-like behavior that I'm *loosely* classifying with the FEAR system because OCD has traditionally been classified with anxiety disorders in the *Diagnostic and Statistical Manual of Mental Disorders* and because people with OCD experience a huge amount of anxiety and fear. However, there's a lot going on in OCD research and the classification may be changing. Dr. Panksepp says obsessive-compulsive thoughts and behaviors could be related to overarousal of the SEEKING system.[25, 26] A lot of people think obsessive behaviors are hardwired survival behaviors such as grooming that have gotten out of control somehow.

Until we know more, I'm going to assume that the cat's very small frontal lobes compared to the rest of its brain have something to do with it. A human being's frontal lobes, which sit behind the forehead, take up 29 percent of the brain, a dog's frontal lobes are 7 percent of its brain, and a cat's frontal lobes are just 3.5 percent.[27, 28] The frontal lobes are important to the *executive functions*: planning, organizing, staying on track, and *changing* actions and plans easily and smoothly. It's that last one that obsessive people and cats have trouble with: They get stuck on one behavior or thought and can't shift to a different one.

Cats tend to be set in their ways, much more so than dogs,

and a lot of times they don't adapt well to new circumstances. I know of one small shelter where they have about forty cats all housed together inside one medium-sized room with cages along the walls, mostly with the doors left open. There's a big island, like a kitchen island, in the middle. The shelter lady tells people not even to try to adopt the adult cats living there because they always end up being brought back. The cats have formed a colony, and the room is the colony's home. They don't want to leave. That doesn't mean you can never adopt an adult cat, but a cat that has spent just a few weeks at a shelter will probably have an easier time adjusting to a new house.

Cats get intensely attached to their homes, too. My aunt had a cat on her ranch who had a terrible time adapting when my aunt moved to a new house five miles away. The cat kept running away and going back to the old house, which was empty. Each time he ran away my aunt would go get him and bring him back until finally the cat stayed. I talked to another lady, though, whose cat never did get used to the new house. The cat kept running away back to the old house until finally the new people who'd moved in said they'd adopt her so she could keep on living where she wanted to.

Cats can be obsessive about their surroundings, too. A man told me that his indoor cat notices every tiny little change in the environment and will stand by a dripping faucet until someone comes and turns it off. Many cats refuse to drink water out of their water bowls. Sometimes this is due to a dirty bowl. They have to drink water straight out of the tap. Someone else told me their cat will only drink water out of a pan they keep under the bathtub faucet because the faucet has a drip.

I think cat obsessions may be one reason why so many cats get into predicaments they can't get back out of. They get mentally stuck. I had to rescue our cat Bee Lee from underneath the dishwasher one time when he couldn't change course and go back the way he came. We had an old-fashioned dishwasher where the whole dishwasher tray pulled out like a big drawer the way a trash compactor pulls out. Someone had left the dishwasher drawer pulled all the way out, and Bee Lee went through the sink cabinet and crawled all the way up under the drawer. Then he just kept trying to go forward to get out into the kitchen even though the space between the floor and the decorative panel at the bottom of the dishwasher was only about an inch high. His paws were sticking out and he was yowling because he couldn't get the rest of his body out.

His SEEKING emotion made him go under the dishwasher to explore the environment. But when he reached the front of the dishwasher, he couldn't put himself in reverse. The sight of the kitchen floor up ahead was "pulling" him forward.

I had to slowly push the drawer in all the way to make him back up to the wall. Once I pushed him most of the way back to the wall he went back under the sink cabinet and got out.

I wouldn't be surprised to learn that a lot of elimination disorders have OCD-like qualities. A cat gets anxious about its litter box or about an invader cat, starts pooping and peeing outside its box, and then can't *stop* pooping and peeing outside its box.

RAGE

In *Animals in Translation* I wrote about "cat explosions" that happen in veterinary clinics when an indoor cat sees a dog for the first time and freaks out. I saw a chart where someone had written, "Assistant was carrying cat down hall when cat exploded." The lady who does my web page, Julie, was bitten down to the bone and had to be on antibiotics for six weeks when her cat saw a dog. In those two cases, the cats probably went from calm to intense FEAR to intense RAGE, because the FEAR system can activate the RAGE system. That's why antianxiety medication can help cure a case of intercat aggression.

Cats can go from calm to terrified or enraged in seconds because they have small frontal lobes, and the frontal lobes are the brain's brakes. They inhibit emotion.

Small frontal lobes also make it hard for a Halloween kitty with all its hair standing on end to calm back down. The cat who attacked his owner because he looked like the bad man who beat his dog had been upset for two hours when his owner came home and he hadn't relaxed at all. It was a really vicious attack, so bad that the owner had to run upstairs to get away from his own cat.

Cats are so violent when their RAGE system is activated, and they stay aroused for so long, that if you have more than one cat in the household you have to be careful never to let them get into a scratching, clawing fight. Catfights are horrible and one bad fight will likely lead to more fights. One summer when I was a child my mother woke up to find our cat Bee Lee in a knock-down, drag-out fight with a tomcat *on her bed.* The tomcat had come in through the window and the two cats were ripping each other apart.

I don't want to scare anyone out of getting more than one cat. Cats need companions and most cats live peacefully with each other. But cat owners should be prepared to handle a fight, because under the wrong circumstances two cats that get along perfectly well can attack each other. The term for this is *redirected aggression.* Redirected aggression happens when a cat is prevented from attacking one cat, and so it attacks another cat or person instead. Dr. Dodman's two cats, a mother and a daughter, had an episode of redirected aggression once. The cats were totally attached to each other and slept curled up together at night. Then one day the mama saw a stranger cat sitting on the deck just outside the screen door. She puffed up into Halloween kitty and attacked her own daughter, presumably because she couldn't get to the other cat. (The cat who attacked his owner was also doing redirected aggression.)

Dr. Dodman separated his two cats as soon as the fight broke out by shooing both of them into another room. Then he shooed the daughter cat into a back staircase, shut the door, and left the cats separated overnight. The next morning they were fine.[29]

If cats that live together get into a serious fight and you can't break it up in time, it's not easy to get the cats back together, though it can be done. If the aggression has been mild, Dr. Debra Horowitz, a cat behavior specialist in St. Louis, Missouri, advises putting a bell on the more aggressive cat's collar. That can break the cycle of fighting because the victim cat can hear the other cat coming and calmly move away.

If the aggression has been more serious, you can reintroduce the cats by putting the more aggressive cat inside a cat carrier and letting the victim cat go into the room. If the aggressive cat starts hissing and spitting, Dr. Horowitz says you should throw a blanket over the carrier. Never put the victim in the carrier, because the aggressive cat will intimidate it.

If those approaches don't work, you might have to start a very gradual desensitization program that can last for many weeks. You should definitely also consider medication for the cats. Dr. Dodman has successfully used BuSpar, an antianxiety medication, even with cats whose fighting had gotten so out of control they didn't respond to months of behavior modification.

We had two cats who didn't like each other when I was growing up: Bee Lee, a neutered male Siamese cat, and Bootsie, a black and white female mongrel. Bee Lee was six years old when we got Bootsie, who was around three. They got into a knock-down, drag-out fight, and after that we kept the peace by having Bee Lee live upstairs and Bootsie live downstairs. Also we built a gate on the front stairs that seemed to help. The cats could easily jump over it but they didn't. They seemed to accept it as a boundary. Eventually they got to where they could tolerate each other. They could be at opposite ends of the living room without fighting. But they never learned to like each other.

If worst comes to worst, and your cats keep fighting, you'll have to find a new home for one of them. Catfights are just too destructive for both them and you to live with.

FEAR and RAGE and Mixed Emotions

I think it's possible wild animals have more complicated emotions — or maybe just more complicated ways of expressing their emotions — than domestic animals because they aren't as neotenized. I say this because human adults have more complicated emotions than human children, so maybe animal adults, which is what wild animals are, have more complicated emotions than neotenized domesticated animals where the adults retain juvenile traits.

By the same reasoning, housecats might have more complicated emotions than dogs because they're not as neotenized as dogs. One sign that cats might have more adultlike emotions is that, according to Dr. Overall, cats can be passive-aggressive. When one cat is mad at another cat but is too scared to fight, it will spray to make a *passive threat.* Or, if a more confident cat threatens a timid cat, the timid cat will spray *after* the confident cat has left. The timid cat isn't going to risk a direct confrontation; it just "says something bad" after the other cat has left.[30]

Bee Lee, when he did not get enough attention, would poop right in the center of Mother's pillow. Mother used to call it "spite potties." A dog wouldn't do that. Researchers are also finding that elimination disorders often go along with aggression problems. The reason aggression and elimination are associated is that cats use spraying, urinating, and defecating to communicate with each other about a conflict. Because a lot of times cats mark *after* the threat is over, their owners don't know what happened. If people could read the smells, they'd have a better chance of finding out what their cats are reacting to.

The PANIC System and the Social Life of Cats

Cats are much more social than people realize, and there are lots of individual differences between cats. There would have to be, or cats wouldn't live and interact with humans. One of the most sociable cats I've ever known was my aunt's tomcat Tomasina, who lived on her guest ranch in Arizona. Tomasina was a regular tiger tabby with black and gray stripes, very friendly and nice, not a scaredy-cat. All the guests could pet him. Tomasina was so sociable he would get in the bathtub with you when it was filled with water. Everybody loved Tomasina.

I've also heard lots of stories of cats figuring out how to get their way with people. A lady I met in Illinois told me about her cat who always wanted to be let out of the house at five in the morning. When she and her husband first got him he would jump on their bed and yowl until someone let him out. She always slept through it, so after a while the cat knew to wake up her husband, not her. He developed a system of sitting on her husband's chest and tapping him politely on the nose.

They were living several hours away from her parents' house, so when they went to visit they would take the cat with them. Apparently the cat figured out that the lady's dad was "the man of the house" there, not her husband, so instead of asking her husband to let him out at dawn he would jump up on the shelf over her parents' bed and knock things down onto her father's head until he woke up. The cat never knocked anything down on her mother's head, just on her dad's.

The social life of cats has been studied in colonies of laboratory cats and, a few times, in free-ranging domestic cats. A cat's social life is somewhat different inside lab colonies versus out in the open, although researchers need more information on how the two compare.

Inside labs, cats form status hierarchies. So far, researchers have been finding that linear dominance hierarchies in captive colonies are based on size. The dominant cat gets to eat first. Usually the dominant cat is the biggest and the oldest, so the dominant cat is mostly going to be male since male cats are three and a half times bigger than the females. But it seems likely that females are sometimes the dominant cat. Cat owners have reported having female cats that dominate the male cats in the house.

We don't know exactly how the dominant cat gets to be dominant. Dominant cats don't beat up the other cats to maintain their position. Instead, the other cats defer to the dominant cat, although nobody understands why the lower-down cats do this. Researchers assume the dominant cat must be doing some kind of threat behavior humans can't perceive. Also, a dominant cat tends to sit on the highest spot in a room, which might be a signal to the lower cats that they have to defer to it. But it could be the other way around, too. The dominant cat gets to sit on the top shelf *because* it's dominant. All cats like to sit up high and the top cat gets the choice spot.

Cats get along very well inside captive cat colonies. You can see this easily in small shelters. To a human, it looks like those cats are living in horribly overcrowded conditions. I visited a shelter like that once. There were cats *everywhere*—dozens of cats in a room the size of an ordinary bedroom. But none of them were fighting, and none of them looked like they had *been* fighting. A lot of them were cuddled up together. These were all adult rescue cats with different life histories. A lot of them had been abandoned, and some of them had probably been abused. But they were living together peacefully in a small, crowded space.

Cats living in laboratory colonies don't fight very much, either. Dr. Overall mentions a time study of a lab colony which found that cats quarreled with each other only 1 percent of the time. That particular colony had seven castrated males in it and one intact male, and they got along fine.

Free-ranging cats are highly social. One reason people have always thought they weren't is probably that cats hunt alone. But cats don't hunt alone because they're loners; they hunt alone because they chase tiny prey animals such as mice and birds that they aren't going to share. Big cats, like lions, hunt together. One lion can't eat a whole wildebeest by itself and can use help bringing it down.

Free-ranging cats naturally form cat colonies, but cat colonies outside labs don't seem to have linear dominance hierarchies. A cat colony can be anywhere from two to fifty-two cats, although researchers have found single cats living on their own, too. Researchers have seen the most interaction between males in the medium-sized colonies.

The female cats in free-ranging colonies spend 40 percent of their time together, usually with their bodies touching. Female kitties look like they're more social than males, but that might be because female cats from all the feline species spend 80 percent of their lives with their kittens or their cubs, so they're almost never alone. Cat owners often say male cats are more affectionate to nonoffspring than female cats, so there may not be a sex difference in sociability.

The most amazing story I read while researching cats was about a cat who helped her sister give birth. First one of the sisters gave birth to three kittens in a nest she had made inside some bales of straw. Eighteen days later her sister joined the nest to give birth to *her* litter. She rolled over on her back, and her sister licked her vulva and helped her deliver all five of her kittens. The midwife-cat licked all of the kittens clean, chewed through the membrane, and bit the umbilical cord in two. This story was reported in 1978, and since that time cat researchers have found that *communal breeding* is normal in farm cats. Mama cats living in communal nests also nurse each other's kittens and help carry them to the new nest if they decide to leave the old one.

The males in the cat colony studies don't fight each other over sex. Dr. Overall mentions a set of pictures one researcher took of a bunch of barnyard tomcats waiting their turn to have sex with a female cat that was in heat. Nobody was fighting anybody; they were just waiting their turn. The male cats in that study were mainly just aggressive to intruders, so I think the tomcat fights people hear are probably between male cats that don't belong to the same colony.

Tomcats don't take care of the mother cat or the babies the same way father wolves do, but that's about all researchers know right now. One farm study found that male cats did take care of kittens, and a man I met who grew up with a lot of cats in Yonkers told me an interesting story about a male cat named Sammy and his kittens.

The man and his mom adopted Sammy and his brother Kwanzaa from a mama cat who lived in the neighborhood. She was a tame alley cat; she didn't have a home. Kwanzaa was a few weeks old, and Sammy was his big brother. Sammy had been a street cat for about a year, and the man told me that even after Sammy started living with him and his mom he still had a "big outdoor life."

One day Sammy met a female cat and went out gallivanting with her. The mom and her son would see Sammy running around with his girlfriend and they'd say, "Sammy's never coming back." Not too long after he started gallivanting with the female cat, Sammy disappeared.

Then, about three or four months later, he came walking up to their apartment building with four kittens beside him. The son told me, "He wanted us to see his babies. It was like, 'Bye, Sharlene. Bye, William. You've been great, but this is what I'm doing now.'" Then he left with his kittens and they never saw him again. He went back to being an alley cat.

A story like that makes me think we don't know enough about the natural social life of cats to say what kind of relationship male cats might have with their kittens. Ethologists believe that male cats have nothing to do with raising kittens. That's probably true, but do male cats sometimes know who their kittens are? We don't know.

Lassie Moments

Cats are social enough that I have now heard several "rescue cat" stories from owners. I've even heard of a cat saving her owner's life. A lady in Texas who has three cats told me that her oldest cat always licks her fingers at night when she wants to be petted. One night the cat licked her fingers while the lady was sleeping. When the lady didn't wake up, the cat started nibbling her fingers and then finally bit her until she was completely awake. It turned out she hadn't turned the gas all the way off on the stove. The lady doesn't know for sure that the cat was trying to warn her about the stove, but on the other hand the cat had never done that before and has never done it since. Since cats hate change, I think it's very possible the cat was trying to alert her owner to something being different in the house that she needed to fix.

The man who told me the story about Sammy also told me a story that sounds like a true "Lassie moment," when a cat defended his owner against people who were threatening him.

The man and his mom had a cat named Bobby Cat who "hated kids to death" and would go hide whenever children came over. The reason he hated kids was that one day, when he was about six months old, he jumped out of their apartment window and fell seven stories to the ground. The boy and his mom weren't home when it happened. The cat didn't get hurt very badly — he didn't even need to go to the vet — but he was probably terrified. When the mom found him he was crouched underneath the building's laundry vent and a bunch of kids were throwing things at him, so he may have associated the fall with the children. After that he hated all children.

One day the boy was home sick from school and somebody knocked on the door. Being only twelve years old, he opened it up without asking who was there, and two boys from the neighborhood pushed their way into the apartment. One was thirteen years old and the other was seventeen, and they'd been taking video games away from middle-school kids.

They went into the boy's bedroom and started going through his games, looking for something they liked. The boy was really upset and scared and kept telling them they had to get out of there or he'd get in trouble with his mom for letting people in while she was gone. But the two boys wouldn't leave.

Suddenly the cat jumped out from wherever he'd been hiding, landed on the dresser, arched his back, and screamed at the boys. The cat was pitch-black and the boys were superstitious about black cats so they took off running. The man who told me the story said they ran so fast they looked like a Scooby-Doo cartoon. Their shadows were still there, but their bodies were gone. Bobby Cat *never* showed his face when kids were around until his owner was scared and possibly in danger. Then he chased the bad guys away. That is a Lassie moment.

The most amazing "Lassie" story I heard, though, was from a lady in Sacramento who keeps lots of animals at her house. She has two cats, a dog, birds, and rabbits, and she also runs a dog-sitting service so they usually have another one or two dogs at the house.

The two cats always slept in their beds in the garage and came and went through a cat door. One night there was a big rainstorm, so the lady decided to bring the cats in for the night. Around one o'clock in the morning, one of the cats, Rainbow, came into the lady's bedroom, jumped up on the bed, and meowed in the lady's face. Rainbow was part Siamese and was really vocal the way Siamese cats are, so the lady ignored her at first. She pushed her off, but the cat jumped back up on the bed and meowed in her face again. The lady pushed her off for the second time, but the cat came right back and meowed in her face.

Finally, after the third time pushing the cat off the bed and having the cat jump back up and meow, the lady got out of bed, picked up the cat, put her out in the hallway, and shut the bedroom door. That didn't help her get back to sleep, though, because the cat just stood outside the door meowing.

Finally the lady got up, opened the door, and looked at the cat. The cat walked down the hall a few steps, then turned around and meowed at the lady. Then the cat walked a couple more steps down the hall and turned around and meowed again. Then the cat did it a third time. The lady said she was just staring at the cat, thinking, "What is her problem?"

The only thing she could think of was that the cat wanted to be let out to the garage so she could sleep in her own bed. So the lady walked down the hallway with the cat, thinking that's where the cat was leading her.

But it turned out the cat didn't want to go to the garage. Instead, she went into the living room and stood by the sliding glass doors. She didn't meow anymore or ask to go out. She just went over to the doors and stayed there. The lady followed the cat and watched the rain for a couple of minutes. She could see that the little creek out behind the house had come up over its banks a ways but didn't think anything of it. The creek had flooded its banks only three times in all the years they'd lived there, and it had never been a big deal when it did. This rainstorm wasn't as bad as those storms had been.

While she was watching the rain she started to worry about the rabbits that were in their hutch outside. The hutch was protected, but the rain was coming down at an angle and she started to think they might get wet. Since she'd always been told you shouldn't let a rabbit get wet and cold, she decided to bring them in.

She went outside and started feeling her way across the yard to the hutch — and when she got to the rabbits she realized that the flooding was much more severe than she'd been able to see from the living room. The creek was already so high it had reached the floor of the hutch.

The rabbits were living inside a huge bale of hay they'd burrowed into for the winter. She had to crouch down in the floodwaters and feel around inside the hay to find them. She finally managed to grab one and pull it out, and by the time she got back to rescue the second rabbit the floodwater was already pouring into the bale. In another few minutes the rabbits would have been dead if the cat hadn't woken the lady and kept on pestering her.

The lady doesn't know whether her cat really was trying to save the rabbits, but the cat didn't bother her again that night after she brought the rabbits in and she never did anything like it again. It's not unlikely that the cat would have a bond with the rabbits. The cat had never been taught that rabbits are prey, and the two daughters in the family often took the rabbits out of their hutch and played with them in the yard, so the cat would have seen them as members of the household.[31] This was a female cat, and female cats take care of all the kittens in the nest, not just their own babies. Maybe the cat's potential for communal breeding was activated by the family's pet rabbits.

Cats Need Companions

Cats need friends and companions to satisfy their social instincts. If everyone in your family is going to be at work or school and you have little time to interact with the cat, you should definitely get two cats so they can keep each other company. Two kittens from the same litter or a mother and her kitten would be best. Other social groupings can work, too, the way they do in cat colonies in the lab or on the farm. But cats are choosy about their friends, so the safest thing is to get cats that are part of the same family.

Many cats that remain alone during the day when the owner works have no behavior problems. Cats tend to be nocturnal, and many of them sleep during the day and become active in the late afternoon or evening. This is similar to the behavior patterns of lions that sleep during the day and hunt when the sun goes down. The cat's natural activity pattern fits with the normal workday, so the owner can play with the cat after work at a time when it will be most active. I remember Bee Lee sleeping for hours during the daytime when the house was full of people. He had a favorite sleeping spot on top of the gas dryer over the pilot light. The metal top was always at a comfortable, warm temperature.

SEEKING

Cats are super-predators. I've had soundmen who were part of a news crew interviewing people in their homes tell me they can't use a windscreen on a microphone because cats will go crazy attacking it. A windscreen is a cover made from really fluffy fake fur that is placed over the mike to filter out the sound of the wind. Cats are built to hunt physically, not just mentally. The position of their canine teeth lets them hold the animal they've caught and dislocate its vertebrae in just one bite. Mama cats start teaching their kittens how to hunt when the kittens are five weeks old by bringing them live prey. Most people have seen cats play with their prey, but nobody knows why they do it. One interesting hypothesis is that cats play with very large or difficult prey, to tire the animal out and reduce its ability to defend itself.[32] Curiosity and learning are also handled by the SEEKING system, and a happy cat has lots of opportunities to explore and learn.

I already mentioned that cats are obsessive about their environments and notice every little change. They also learn from their environments. Cats probably do a lot of social learning. One major study showed that cats learn a new behavior faster when they watch other cats do the behavior than when the experimenter teaches them how to do it using reinforcers,[33] and another study found that cats learned how to escape from a box faster when they watched another "cat student" *learning* how to escape than when they watched a "cat expert" who already knew how to escape get out of the box.[34] Karen Pryor has a great story about one of their cats watching her daughter use bits of ham to train her dog to sit in the rocking chair and rock it back and forth. As soon as the dog left the room, the cat jumped up in the rocking chair, rocked it back and forth, and looked for his piece of ham.

Karen Pryor thinks that cats' reliance on social learning explains why they get stuck in trees. Climbing up a tree is natural; cats don't have to learn how to do it. But to climb back down a tree they have to go *backward* because their claws are curved. Going backward is probably something cats have to learn from their mother, and most pet cats are taken away from their mother before she can teach them. So they run up trees and then don't know what to do to get back down.

I think she's probably right about this because I've never heard of a barn cat getting stuck in a tree, only housecats. Barn cats are tame cats that live on farms and breed freely. The reason barn cats are tame is that the tame mama cat has her kittens where the farm family can find them, so the kittens start getting socialized to humans as soon as they're born. Since the kittens are going to live in the barn, too, nobody takes them away from the mama. Very likely the kittens learn all the lessons a mama cat would naturally teach them. One lesson is probably how to climb backward out of a tree.

The most amazing story I've heard about a cat learning how to do something by watching others do it was a cat who taught himself how to use the toilet. I had heard stories about cats using the toilet instead of a litter box, but I never knew whether to believe them until I met a lady who told me about her friend's experience. Her friend married a divorced man who had two young boys who stayed with him on weekends. The lady didn't have any kids, just a cat.

Not too long after they all moved in together, my friend started finding that someone had used the toilet and hadn't flushed. Sometimes the seat was wet, too. She didn't want to say anything but finally she told her new husband to please ask the children to put the seat up when they used the toilet and be sure to flush. The husband did, and the kids said it wasn't them. Of course she thought it had to be them so the house was kind of tense until one day she walked in on the cat using the toilet to poop. She said the only thing the cat hadn't taught himself to do was to flush.

Another lady told me that her cat almost *did* teach himself how to flush. He wasn't going to the bathroom in the toilet, but he loved to watch the water swish around the bowl whenever somebody flushed. So he started trying to flush the toilet himself, but he couldn't figure it out. Apparently the cat had also noticed that whenever anybody put toilet paper in the toilet the water would swirl. So after he couldn't learn to flush he started batting lots of toilet paper into the bowl. I guess he thought maybe it was the paper that made the water swirl around.

After I heard those stories I looked on YouTube and I found twenty videos of cats using a toilet. I don't know how many of those cats figured out how to do it on their own the way this lady's cat did. Probably a lot of them were trained with a commercial toilet-training kit for cats. The kit is like a kitty litter tray that fits inside the toilet underneath the seat. At first the cat uses it as a kitty litter box, then at some point starts using the toilet without the litter tray. But it's still pretty amazing to watch a cat using a human's toilet.

All of that exploratory learning behavior is motivated by SEEKING. Cats are very curious animals because of their predatory nature. That's probably why we have the saying "Curiosity killed the cat." Curiosity is SEEKING. The best example of cat curiosity I've seen myself was my aunt's tomcat, Tomasina. One night we walked down to the horse barn and found a giant bull snake right outside the door. It was about six feet long, an inch and a half in diameter, all coiled up and hissing. Bull snakes aren't poisonous and don't bite, but they do a good imitation of a rattlesnake when they're threatened.

And there was Tomasina, five feet away, crouched down watching this big bull snake and keeping his distance. He wasn't in a head-forward stalking pose. It was pure curiosity. He wanted to see what the snake was doing. I took a picture of Tomasina watching the snake with my little flash camera.

How to Turn on PLAY and SEEKING by Training Your Cat

The key to animal welfare is to keep the positive emotion systems such as PLAY and SEEKING turned on and to keep the negative emotion systems — RAGE, FEAR, and PANIC — turned off as much as possible. Turning on a cat's SEEKING system for PLAY is easy. Cats like anything that moves because cats are hunters, and a hunter's brain is triggered by movement. You just have to keep them supplied with toys that move. One thing you have to be careful about is never to let a kitten play-hunt by jumping on your hand. That will be dangerous when the kitten grows up. Use a toy on a string or the little feather duster wand I described earlier to play with your kitten so that your hand does not get scratched.

Outdoor cats probably don't need any help turning on their SEEKING system. They can hunt and explore as much as they like. But a lot of people think cats should be kept indoors for their own safety and also so they don't kill a lot of songbirds. I've seen cats do very well indoors, especially when the owner has more than one cat or the cat has lived indoors its entire life. However, cats are genetically adapted to living free out of doors. If you live in a safe country area, you should probably think about having an indoor/outdoor cat.

If you're going to have an indoor cat, you have to think about giving it mental stimulation. In an interview Dr. Overall explained, "People don't see cats as really bright, inherently cognitive individuals. They forget the most critical need, which to me is the intellectual one. I think we haven't given cats or dogs the credit they deserve for their cognitive capabilities. I think we've got an epidemic of understimulated cats whose intellectual needs aren't being met."[35] You can't lock up a lone cat inside a bedroom and go off to work twelve hours a day. That kind of life is no better than what zoos used to do with their captive animals.

Dr. Overall recommends that owners give their cats mazes, cat trees, food puzzles, and screened-in areas outdoors.[36] These things lower a cat's stress level by giving it choices.

I agree with her advice. I also recommend buying some kind of clicker[37] and a book on clicker training to train your cat, because learning new things stimulates a cat's mind the same way learning stimulates a person's mind. Using clicker training to teach a cat new things stimulates its SEEKING system.

Before I talk about how training works, I want to say something about why you need a clicker or something that works like a clicker. You need a clicker for the simple reason that it's extremely difficult to train a cat using just your voice. Amateur trainers find clicker training easier with any animal, but many dog owners can train their dogs in the basic commands — "sit," "stay," and "come" — using purely social rewards such as praise and stroking. It's different with cats. Cats have to be motivated to perform by using food as a reward. Cats respond well to clicker training because they associate the click with getting a treat. People have always thought cats weren't trainable because they did not respond well to purely social rewards. But it turns out that they are highly trainable when you use a clicker. People are even starting to talk about training cats to work as service animals.[38]

Cats have to have food motivators for several reasons. First of all, a dog is socially motivated to please its trainer because of its larger frontal cortex. Large frontal cortices evolved in animals to make more complex social behavior possible. A second reason is that dogs are more fully domesticated than cats, so they read our faces and bodies better than cats do. A dog can detect the tiny micro-changes in your posture and facial expression that happen *before* you say, "Good dog!" I'm sure dogs quickly learn to associate those micro-changes with the words "Good dog!" and the food reward, so they're getting a very rapid signal-before-the-signal that we humans don't realize we're giving out. That's important because speed and timing are extremely important in animal training. To be a good trainer you have to be a good signal giver, and to be a good signal giver you have to be fast. A professional trainer gives the click or the voice reward (or whatever he's using) the instant the animal does something the trainer wants it to do, and the animal has to read the signal that instant, too. But even for the best of trainers, words are slow — slower than clicks or whistles. That doesn't matter as much with dogs because they are probably reading the signal-before-the-signal.

I think it's possible the click gives cats the same super-fast signal that micro-changes in expression and posture give dogs. Cats may be able to *perceive* tiny micro-changes in a person's face or posture, but they haven't evolved to read people, and they aren't motivated to scrutinize their owners for signs. You know a cat is going to hear a click. You don't know it's going to pick up on tiny movements in your face or posture.

The third reason why clickers work for cats is that it's extremely easy to use pure positive reinforcement using the clicker. Remember, cats don't respond to punishment or negative reinforcement. When you use a clicker to train an animal, you start the first session by teaching your cat that "click means treat." You click, then you give the animal a treat. Then you click again and treat. You only have to do this two or three times before the animal learns that the click means food. Trainers call this *charging up the clicker.*

Psychologists call it *classical conditioning.* Classical conditioning is the kind of learning Pavlov's dogs did. Every time they heard a bell they got food, and pretty soon they started to salivate as soon as they heard the bell. Clicker-trained animals probably salivate when they hear the click. The click has become a reward, because the click means food. (Even though the click becomes a reward you can't just stop giving the real rewards. You don't have to give the food treat every time you click the animal, but if you *never* give the food treat the click will stop being a reward.)

As soon as you see that your cat knows "click means food," you switch from classical conditioning to *operant conditioning,* which is teaching your cat to do something in order to get the reward. You start with something very simple the cat is already doing, like looking at you or sniffing something you've put down in front of him. Very quickly your cat will learn that "click means keep doing what you're doing," and he will purposely look at you or sniff the object in order to get the click and the food. As soon as he learns that behavior you start clicking a different simple behavior because you want your cat to "learn to learn." In the first session your goal is to teach him: "I need to figure out what to do to get my owner to click." Karen Pryor says that what you're doing is teaching the animal to think that he can train you by experimenting with a lot of different behaviors until he finds the one that makes you click.[39]

With clicker training all of this is done using only positive reinforcement. There's certainly no punishment, but there's also no *negative reinforcement,* meaning the animal isn't learning new behaviors in order to get you to *stop* doing something he finds unpleasant, such as dragging him by the leash. To train a cat to walk happily on a leash, you would use the clicker to train him to touch a target, such as a wooden spoon. Then you would walk him on the leash with the target in front of his nose so he'll touch it in order to get the click and the food. When he learns that, you fade back the spoon (take it away gradually) and click the cat for walking calmly on the leash.

I recently watched a cat going through an obstacle course. His trainer had attached a clicker to the handle of a long wooden spoon and put food in the bowl of the spoon. He was using the "clicker spoon" to lead the cat through the obstacle course, giving him clicks and food rewards every time he did something right.

The big cats at the zoos are happier and less bored when clicker training is used to train them to leap from one stand to another. At the San Diego Zoo, cheetahs and clouded leopards participate in demonstrations for the public.[40] I have seen miserable clouded leopards that refused to come out of their enclosures. They would probably be less fearful if they were put in a clicker training program that would turn on their SEEKING system.

Cats need a strict program of positive reinforcement — rewards — to learn, and humans use their voices for every kind of communication, good, bad, and in between. A good trainer could associate a word or a tongue click with treats, but I think that a physical clicker or whistle helps regular people to stay completely positive. Using negatives comes naturally to people and animals because of our core emotions. If you're trying to train a cat and the cat doesn't learn, you're likely to feel frustration, which comes from the RAGE system. Being negative is natural, and being 100 percent positive takes work.

Why the Click Turns on the SEEKING System

When I was in college, I was taught that the reason Pavlov's dogs salivated when they heard the bell was that the bell had turned into a reinforcer. The food was a *primary reinforcer,* meaning something that is naturally, biologically rewarding. The bell was a *secondary reinforcer* because it wasn't rewarding in and of itself, but had become reinforcing because it was associated with a primary reinforcer. Money and good grades are both secondary reinforcers for humans.

But now that researchers understand dopamine and the SEEKING system better, the way we think about rewards is changing. What's rewarding about rewards isn't so much the reward item itself, but the time you spend looking forward to it. In some ways, chasing after things is more fun than actually getting them.

That's why a charged-up clicker turns on the SEEKING system. The click isn't the reward; the click is the signal that a reward is about to come. The sound of the click turns on the SEEKING emotion and the animal goes into a very pleasurable state of eager anticipation.

To meet a cat's social needs you need to keep the FEAR, RAGE, and PANIC systems turned off as much as possible. You do that by providing friendly, positive companionship, either with you and your family or with another cat.

To meet a cat's intellectual needs you have to turn on the SEEKING system. Cat toys and clicker training are the best ways we have to do that for a cat.

4 Horses

THE HORSE IS AN animal that survives in the wild by fleeing and kicking at predators that are attacking it. A horse is all about flight, and fear is the dominant emotion.

Horses are a prey species herbivorous grazing animal. *Prey species* means that it is killed and eaten by predator animals such as wolves or lions. *Herbivorous grazing animal* means that it's a vegetarian that gets its food by grazing. There are two basic behavior types in these animals: cattle and sheep, which bunch together for safety, and horses, deer, and antelopes, which use flight for survival — basically, the bunchers and the flee-ers. That's an important distinction to know because flight animals are more skittish and startle more easily than prey animals that bunch, which makes them easy to traumatize.[1] Horses are much more flighty than cows, sheep, or goats. What sets off both bunching and fleeing is *novel rapid movement*, meaning a sudden, unfamiliar, and fast movement the animal isn't expecting or hasn't seen before. In a grazing prey animal, vision is more important for survival in the wild than all the other senses, because rapid movement is a signal that danger is imminent. You might have seen photographs of antelope herds peacefully grazing while a herd of lions suns itself off in the distance. The reason those antelopes aren't fleeing is that the lions aren't stalking them, and the reason the antelopes know the lions aren't stalking them is that they can see that the lions are lying down. Their sense of vision tells them when a predator is dangerous and when it's not.

If a horse is cornered and can't flee he'll turn around and kick the predator with his hind legs. Horses are smart, with a big frontal cortex. I'm sure you've seen horses standing alongside each other, with each horse's head next to the other horse's butt, and flicking flies from each other's faces. I've never seen cattle figure out how to flick flies off each other, though maybe because their tails are hard and ropy they prefer not getting whacked in the face by another cow's tail.

Wild mustangs out on the western ranges live in small groups. A stallion lives with four to six mares and the other stallions live in little bachelor groups. Since they survive through flight, they don't need a big group to protect themselves. New baby foals can run and flee almost as soon as they're born.

The horse is a bit like the dog; it wants to please you. Expert riders have been trying to describe the horse's social connection to its rider for centuries. The oldest riding manual we have is *On Horsemanship,* written by Xenophon sometime between 365 and 345 BC. Xenophon writes, "What a horse does under compulsion he does blindly, and his performance is no more beautiful than would be that of a ballet-dancer taught by whip and goad. The performances of horse or man so treated would seem to be displays of clumsy gestures rather than of grace and beauty. What we need is that the horse should of his own accord exhibit his finest airs and paces at set signals."[2]

Horses and the FEAR system

Horses aren't naturally tame when they're born, and they're extremely high-fear. You have to socialize a foal to humans from the day it's born. Since the 1990s a technique called *foal imprint-*

ing has been popular, which involves intensive handling of the newborn foal. Some behaviorists are concerned that foal imprinting may stress the foal too much because the newborn foal has to be forcefully restrained while the procedure is being done.[3] I don't know whether foals are stressed, because some forms of pressure restraint, like full-body restraint for aggressive dogs, are calming. But I've always thought foal imprinting was too rough, and so far the studies on it haven't shown lasting effects.[4] I was glad to see new research come out in 2005 showing that the best way to habituate a newborn foal to humans is to brush the mare fifteen minutes a day for five days after the foal's birth and have the foal watch. The foal learns that his mother likes to be brushed, so he is more accepting of humans, too. In the study, foals who watched their mamas being brushed were much friendlier to humans at one year of age than foals in the control group where a person just stood next to the mare. They accepted saddle pads being put on their backs more quickly and they were friendlier to strangers, too.[5]

Rough Training Methods May Ruin a Horse

Unfortunately there are still some trainers who use an old-fashioned rough method they call *sacking out* a horse. It is still common on western ranches. Instead of slowly and gently habituating the horse to new things, they throw them at the horse all at once. When a trainer sacks out a horse, he takes a yearling, puts a really strong halter on its head, and ties it up to a post. Then the trainer throws stuff at the horse: tin cans, blankets, pieces of plastic — anything he has. All horses initially react by pulling back and trying to get away. Some horses panic, fall down, and become traumatized, and others habituate and learn to tolerate all the novel things being thrown at them.

The goal of sacking out a horse is to make the horse not be afraid of new things. The handler just keeps throwing stuff at the horse and makes the horse fight it out against the post until the horse gives up in exhaustion or habituates. It's similar to *flooding,* a behavioral technique where you expose a person to a huge amount of the thing he's afraid of all at once. If he's afraid of elevators, you shove him inside an elevator instead of trying to slowly desensitize him.

It's a horrible thing to do to a horse. The hot-blooded Arab-type horse, you traumatize and wreck. I think you get a post-traumatic stress disorder (PTSD) effect, similar to a traumatized war veteran. A horse treated this way can't be ridden. A less reactive horse, like a quarter horse from a bloodline that's really calm, can take it, but it's still a bad thing to do. The terms *warm-blooded* and *cold-blooded,* by the way, refer to a horse's temperament, body shape, and sometimes its breed, not to a blood type. The hot bloods, such as Arabs and Thoroughbreds, have slim bodies with slender legs, and cold bloods, such as draft horses, are stocky with heavier bones. Measurements of the diameter of the leg bones show that the most flighty animals have thinner bones.[6] No one should abuse any horse regardless of whether the abuse works or not. I've noticed that rough trainers prefer to work with horses that are calmer and less reactive.

Bombproofing Your Horse

Successful training of horses involves habituating them to strange, novel sensations, including the bridle, the saddle, and having a rider. If new sensations are introduced too rapidly, a horse can panic, and the best trainers never produce panic reactions of bucking, kicking, or rearing in their horses. New sensations have to be introduced more slowly in horses with high-strung, nervous genetics.

Proper habituation of horses to the novel stimuli they'll encounter in their leisure riding or working lives takes a long time and has to be tailored to the life the horse is going to lead. A riding horse needs to be habituated to different stimuli from a police horse or a stunt horse.

In most parts of the country, the most important fears to inoculate your horse against are fears of bicycles, flags, and balloons, because these things are really common at horse shows and parades. Bikes are scary because they are rapidly moving and silent and they sneak up on a horse. Flags and balloons frighten horses because they make rapid erratic movements and are high-contrast light and dark, which also frightens a lot of animals. Also, Mylar balloons make weird sounds sort of like a fire burning.

Habituating a horse to bikes is easy. Just wheel the bike up to your horse and let him get used to it. Then ride the bike slowly for a short distance, gradually riding faster and farther as the horse habituates. New things will be less scary if you get the horse's SEEKING system turned on. Wendy Jessley, a horse trainer, gets horses accustomed to new things by having the horse follow them. She suggests slowly pushing the bike away from the horse and then gently leading the horse to follow it at a safe distance. The best way to get a horse used to flags is just to tie one to the fence and let him walk up and explore it on his own. You can do the same with balloons. If you plan to show your horse or ride him in many different places, he must be habituated to a variety of sights and sounds, including other animals he may encounter such as llamas. He should also be introduced to many common objects such as chairs, baby strollers, and children's large toys. When a new thing is introduced, you need to be a good observer to determine which sense is triggering fear. Most of the time it is vision, but sometimes sound or smell is the trigger. I rode my horse King in a costume show where he went as a space-age horse and I was dressed as an astronaut. I carefully introduced King to the costume before the show. He came up and sniffed the tin-foil-covered helmet when

I held it still, but he jumped away when the helmet moved. I finally figured out that he was afraid of a weird little sound made by a jiggling wire on the helmet. For King, the sound of the wire was more frightening than the sight of the helmet. I solved the problem by removing the wire, because it would have taken a long time to habituate him to it.

Horses get more easily spooked by novel sounds and sights, not so much by novel smells. When researchers exposed horses to novel sights, sounds, and smells — traffic cone, white noise, and eucalyptus oil — their heart rates went up for the traffic cone and the white noise but not the oil.[7] That's because horses don't usually use smell to detect danger, so they aren't hypersensitive to strange smells. So this is one less thing to worry about unless a horse has been abused by a person with an odd smell. I heard about one horse who feared people with alcohol on their breath, but that's unusual. In most situations the horse is more likely to associate the abuse with a visual or auditory stimulus. Some common examples are a bearded man or a high-pitched voice.

Hyper-specific Fear Memories — and How to Handle Them

In any habituation program, you have to expose the horse to all the different scary things he's likely to see. You can't just expose him to one or two scary things and teach him the general principle: *Don't be scared of novel stimuli.* You have to habituate him to each scary thing separately.

That's because animals — and autistic people — are supersensitive to sensory-based detail. They are *hyper-specific.* In *Animals in Translation,* I wrote about a horse who was afraid of black cowboy hats but not afraid of a white cowboy hat or a baseball cap.

People have always known horses were sensitive to detail, but they haven't always interpreted this the right way. For example, horses sometimes startle when they see the same object from a different angle. This can happen in the arena or on a trail when you ride the horse in one direction and then back again. An object that didn't scare the horse on the trip out can startle him when he comes back the other way.

The explanation for this used to be that the horse brain doesn't transfer information between the eyes. Horses have eyes on opposite sides of their face, so it was thought that a horse that doesn't startle at the sight of a flag in an arena on the way across but does startle on the way back is actually seeing the flag for the first time *twice,* once with each eye. But anatomical and behavioral research has proved that the horse brain does transfer information from one eye to the other. I believe that the real explanation for why a horse can startle when it sees an object from a different angle is that the object looks different and therefore becomes a brand-new, scary thing. That's true for me. If I pass a barn on my way to an appointment, that barn looks like a different object to me when I'm driving back from my appointment because I am seeing it from a different angle.

Dr. Evelyn Hanggi did a study to see whether this could be the explanation and found out that it could. When she rotated children's toys into different positions, the horses she tested recognized the toys in some rotations but not in others. Even though the horses were looking at the same toy they'd been looking at just a few minutes earlier, they didn't realize it was the same.[8]

Almost all of the consultations I do with horse owners are about hyper-specific fear memories their horse has and what to do about them. If the owners know the horse's handling history, we can usually figure out what the problem is. Recently I talked to a person whose horse was deathly afraid of long-handled tools. The owner was able to trace this back to an accident in the crosstie[9] when the horse flipped over and fell down, pulling his lead rope taut across his chest. It was a thick blue rope, and when it was stretched tight it looked like a broom handle to the horse. So now he was scared of a blue-handled squeegee, and he had also panicked when he saw a broom.

This is a case of visual hyper-specificity because the horse wasn't thinking about the *meaning* of "rope" or "tool handle," but about the super-specific visual category of "things that are long, straight, three-quarters of an inch in diameter, and probably blue." No one else has that category, only that one horse. I told the owners just to keep him away from tools with long handles unless he's being ridden by his best-friend rider who can calm him down. If some kid gets on him and the horse sees a rake leaning up against the wall of the barn, the horse is going to freak out. With fear memories, it's always best just to get rid of the thing the horse is scared of if you can. I'll talk about extinguishing fear memories later on.

Another owner I talked to had a horse who was scared to death of riding crops if she held one in her hand while she was riding him. If she held the riding crop in her hand while she was standing on the ground, he didn't mind. This is another example of a fear being hyper-specific. The riding crop triggered a fear memory only when it was in a person's hand while the person was on the horse's back. That's because nothing bad had ever happened when a person standing on the ground was holding a riding crop. In another case, a horse was afraid of naked white saddle pads. If a saddle partially covered the pad, the horse tolerated it, but a naked white pad either on a fence or on another horse's back was scary. It is likely that the poor horse had been roughly sacked out with a white saddle pad. Dark pads had no effect.

Sometimes problems due to hyper-specific fear memories are easy to avoid. For instance, a couple asked me about their horse who would load only on the left-hand side of the trailer, not the right. They didn't know her history, but she had probably had some kind of accident on the right side. I told them just to load her on the left-hand side and not to worry about it.

Another couple got in touch with me about a mare they had bought to train as a carriage horse. She was quiet and calm when the carriage harness was put on, but as soon as she pulled forward and felt the pressure of the harness on her back, she went berserk. When they looked into her history, they found out she'd been used for collecting urine to manufacture the drug Premarin. To collect the urine, ranchers keep pregnant mares locked up inside a stable all winter with a rubber collection cup shaped like a jai alai scoop attached to their hindquarters. They harness the horses to the stalls so they don't knock the collection cups off.

The rancher who owned this horse had made his harnesses out of strips of rubber from a tractor tire inner tube. That made the harness stretchy so the horse could lie down. One day the mare had gotten loose and walked twenty feet away, which stretched the inner-tube harness out behind her until finally it snapped like a giant rubber band and whapped her on the butt. When her new owners harnessed her to the carriage, the pressure of the straps woke up her hyper-specific fear memory of getting whapped. Since the mare was a good riding horse, I told the couple just to ride her and not try to train her to pull a carriage.

Different Causes of Fear Memories

A fear memory can have two causes. The first is a past abusive experience and the other is introducing a new thing or a new sensation too quickly. It's best to prevent fear memories from forming in the first place because a bad fear memory is very difficult to completely correct. A horse's first experience with a trailer, shoeing, or new equipment should be very positive. A bad first experience is more likely to create a fear memory. If possible, teach your horse to go in a long trailer, which is less scary to enter, before introducing him to a smaller two-horse trailer. Horses can be hyper-specific about trailer fears. One horse I know of entered a trailer very easily but when he was unloaded, he bolted out like a rocket. In the past he had banged his head backing out of the trailer so now he backed out quickly because he was trying to get out before the trailer could hit him.

Fear memories that are associated with bits or other tack can be caused by abuse or by introducing the item too rapidly so that the horse never habituated to it. Fear memories associated with tack are also hyper-specific. A thirteen-year-old girl in 4-H had been showing her horse for three years without a problem. The horse was perfect. Then all of a sudden the horse went crazy and the family eventually had to sell him. When they heard me lecture about hyper-specific fear pictures inside horses' heads, they remembered that right before the horse broke down, the riding instructor had switched everyone to a snaffle bit.

I'm sure that's what set the horse off. To a person, a bit is a bit, but to a horse, a jointed snaffle bit is totally different from a solid, one-piece bit. I say to people, "When you go home, hold the one-piece bit in one hand and the snaffle bit in the other. If you pay attention, you can see that they feel totally different." I've talked to four or five different people now whose horses were obviously acting up because of a particular kind of bit. The horses had been abused by an owner or rider using that bit and associated abuse with the feeling of the bit. In every case changing to a different bit totally fixed the horses. The 4-H girl's parents wouldn't have had to sell their horse if they'd realized the new bit was the problem. But they didn't make the connection until they heard my talk.

A common problem that is created by introducing new sensations too quickly is a horse that bucks when he changes gait. He does that because the saddle feels different at each gait. If he's been habituated to the saddle at only a walk and a trot, when he moves to a canter the saddle suddenly feels like a novel stimulus.

My horse Sizzler had that problem. My aunt bought Sizzler for me when I was in high school, but he was too dangerous for me to ride because he would buck when he went from a trot to a canter. If I'd known then what I know today, I would have started his training over again and replaced his Western saddle with an English saddle to provide a totally different saddle "feeling picture." Then I would have gradually gotten him used to what the new saddle felt like at each gait. But I was only in high school and I didn't know anything about hyper-specificity, so my aunt had to sell Sizzler back to the dealer.

Behavioral Signs of Fear

It is really important to recognize the behavioral and physical signs of fear. A fearful horse switches his tail. As he becomes more scared, the tail moves faster. Other signs are a high head, sweating when there is little physical exertion, and quivering skin. A really frightened horse gets bugged-out eyes and the whites show. When a horse is being introduced to any new procedure such as loading on a trailer or picking up his feet, training sessions should be kept short and ended before fear escalates into an explosion that can form a bad fear memory. When a few tail switches start, end the training session with the horse doing something right. If a horse gets really agitated during shoeing or a veterinary procedure, the best thing to do is to let him calm down for thirty minutes. Recently I talked to a veterinary technician who tried this with a horse who went ballistic at her clinic. The veterinarian thought the horse would need tranquilizers, but letting him calm down was all he needed.

A common mistake people make is mixing up fear and aggression. Most behavior problems that occur during handling, veterinary procedures, loading, and riding are caused by fear or pain — not aggression. The worst thing that can be done to a frightened horse is to punish him by hitting or yelling. Frightening or painful punishment makes fear worse.

Diagnosing Fear and Pain-Based Behavior

It's very important for horse owners to understand the nature of the horse's FEAR system because many behavior problems in horses are caused by fear. However, sometimes the problem is physical and the behavior problem is caused by pain. When a horse has a behavior problem, you first want to rule out painful medical conditions such as abscessed teeth, injuries to the mouth, saddle sores, or lameness. You should also check to make sure the tack isn't pinching the horse anywhere and that nothing else is hurting him. A horse should be carefully examined by both a veterinarian and a farrier. A single misplaced nail in a shoe can make a horse sore-footed, and a stone stuck in his foot can make him jumpy. If pain can be ruled out, then the behavior problem is probably caused by fear.

Learning Not to Be Afraid

To lower the amount of fear horses experience, the single most important thing anyone can do is to prevent fear memories from developing in the first place. Fear memories are permanent in all animals, including humans. Sometimes a fear can be *extinguished*, but *fear extinction* isn't forgetting; it's new learning.

For example, if you fall off your bicycle when you're learning to ride a bike, you'll never forget the fear that it caused. But after you ride your bike a few times without falling off, you learn that bike riding can be safe, too. The fear memory doesn't go away, but a stronger "fun memory" gets created that *opposes* the fear memory. However, the minute you ride over a big branch in the road or your bike skids on a slippery spot, your fear comes right back. I like to use computer terminology to explain this because it is easy to understand. New learning can *close* the "fear file" but the file is never deleted from the horse's memory. Sometimes the "fear file" keeps popping back open and can never be completely extinguished.

To extinguish a fear in a horse you carefully expose him to very small doses of the thing he's afraid of and then continue to increase the dose until the horse doesn't act afraid anymore. You can also put a horse on a *counter-conditioning* program, where you associate the scary thing with a good thing such as a food treat. If a horse is afraid of a trailer, feeding yummy treats in the trailer may help him get over his fear.

Horses are so high-fear that it's harder to extinguish fears in a horse than in a human. It may be almost impossible to extinguish a severe fear memory in a high-strung horse like an Arab. The horse trainer Cherry Hill gives a list of bad habits in her book *How to Think Like a Horse*, which she divides into curable versus incurable or close to incurable. The incurable bad habits are all fear-based behaviors such as rearing and striking. (*Striking* means the horse takes a swipe at a person with a front leg.) Another fear-based behavior that is difficult to get rid of is constant tail swishing and tail *wringing*. Fancy dressage horses often repeatedly swish their tails when they do the most complex movements. This is usually the result of aversive training methods that were used to teach those movements.

Curable bad habits are things like balking and not cooperating with the farrier. There might be an element of fear motivating these behaviors, but it's usually not intense.

The RAGE system

FEAR is the main emotional system that causes behavior problems in horses because horses are much higher-fear than many other species. However, there are some situations where frustration, which is a mild form of RAGE, may cause behavior problems. I have seen horses worked for so many hours circling in a round pen that they get bored and frustrated. A round pen is great for activities such as initial habituation to wearing a saddle. A person standing on the ground can easily see the first sign of fear before the horse tries to buck the saddle off. The trainer can then end the session before the horse explodes. However, if a horse is forced to go around and around in the same circle for hours, the RAGE system may start to get activated.

Another thing that may frustrate a horse is being confined to a box stall with little opportunity for exercise. I think many aggression problems in stallions may stem from frustration and rage caused by being housed alone with little exercise. Horses with a low-fear temperament could also get their RAGE system turned on if they are severely abused.

You can force low-fear horses to obey you by hurting them physically, but those horses are going to have a lot of suppressed rage, and if you turn your back they're going to get you. I read a horrible book about horse-training techniques in the Middle Ages. The book said people in those times beat horses into submission. The author also described several instances of a horse viciously attacking a person. There was one description of a horse tearing a guy's guts out. The only horse that's going to tear a person's intestines out is a horse that has been very badly abused. No one should teach an animal to hate humans.

The PANIC System and the Social Needs of Herd Animals

Horses are herd animals with strong social needs. You can't keep a horse locked up alone in a stall all the time, the way some breeders have been doing. Horses need companions so badly that owners sometimes place ads looking for a companion animal to buy for their horses.

If horses do have to be alone inside a stall, they should have a mirror so they can see themselves,[10] and it's best if they can have stalls that allow visual contact with the other horses in the stable.[11]

However, just being able to see other horses isn't enough. This is a mistake a lot of people make. Living in separate stalls isn't like living together in a group because most horses in stalls can't do social grooming behaviors. When horses groom each other, their heart rates go down. If horses have to be locked up in stalls, stable owners need to build open-bar walls between the stalls so the horses can touch and groom each other.

Giving a horse a good social life is important for training and riding, too. One study compared young stallions housed alone to stallions housed in groups of three until age two. The horses housed in social groups were a lot easier to train, with less biting and kicking at the trainer.[12]

The horse's sociability is what made it an animal that could be domesticated in the first place. Out of 148 large mammals weighing over a hundred pounds that might have been domesticated by people, only fourteen actually were domesticated, although people tried to domesticate a lot more than that. Out of those fourteen, only five became important in every part of the world: cows, pigs, sheep, goats, and horses.[13] One of the main things that makes a wild animal a "candidate" for domestication[14] is that the animal has to be highly sociable by nature. No large solitary species has ever been domesticated. All of the fourteen are herd animals with clear dominance hierarchies that can transfer the dominant position to a person and follow the person instead of another animal in the herd.

The big animals that were domesticated had one other quality in common, which was that they lived in overlapping home ranges. A lot of herd animals police their territories, and some others have overlapping territories most of the year but not during mating season. Wild cattle and horses, though, don't mind sharing ranges any time of the year. That makes it possible to put different groups together inside the same pen or pasture. The domesticated animals are peaceable creatures compared to the animals that humans didn't domesticate. For example, the main reason zebras never got domesticated is that they're ultra-high-fear. Zebras may bite people and not let go. They injure more people in zoos than the tigers do.[15]

The sociability of horses gives them a desire to please their human owners. When a horse has a good relationship with his rider, he has a built-in, natural desire to cooperate and follow his rider's lead. He can also get reassured by his best-friend rider that everything's OK when he sees a scary thing.

Reasons to Pay Attention to Emotions

The biggest challenge of horse welfare is preventing behavior problems. Careful habituation is required for horses because they are more flighty compared to dogs that perform service jobs such as guide dogs for the blind or police dogs. Horseback riding is dangerous even with a very well-trained horse. One study of horseback-riding injuries in England found that riding horses was twenty times more dangerous than riding motorcycles. Another study broke the risk down by type of riding:

- one injury for every hundred hours of leisure riding
- one injury for every five hours of amateur racing over jumps
- one injury for *each hour* of cross-country eventing[16]
  (Eventing is a two- to three-day competition that includes dressage, cross-country riding, and show jumping. Christopher Reeve's fall happened during the cross-country portion of an eventing competition, which is the most dangerous.)

A service dog needs months of training, but with regular dogs, if the training doesn't work out it is usually OK. A Labrador retriever that fails the training to become a service dog can be given to a family that wants a pet. Horses are less likely to get kept exclusively as pets because it's too expensive. In the West, in Colorado, where you have a neighbor who lets you rent some pasture, that would be $3,000 a year just to graze the horse. My assistant, Mark, has five horses that he grazes for free on high-quality pasture I own, and he's spending on average about $1,000 a year for each horse for vet bills and hay and feed in the winter. Since he is a farrier, he does all shoeing and hoof trimming himself. This saves a lot of money. If you don't own a farm or ranch yourself, $4,000 a year is the lowest amount you can spend to house one horse. If you live in a city, keeping a horse is much more expensive.

If an owner has a problem horse he can't ride, he may try to sell the horse to someone else who might be able to rehabituate the horse. If he can't find a new owner, he may just sell the horse at an auction. One study of around three thousand non-racing horses in France found that 66.4 percent died between the ages of two and seven years.[17] Most of them were put down because of bad behavior. I don't know what the figures are for America, but I know they're high. I have survey data from ten grade horse auctions in ten different states showing that 47 percent of all the horses in the sales were in sound condition for riding.[18] Grade horses are mixed-breed horses or horses whose parentage isn't documented, and a grade horse auction is like a used-car lot. The original owners of the horses aren't making a profit by sending them to auction; they're basically just trying to unload the horses, either because they have behavior problems, which is usually the case, or because the owners can't afford to keep a horse anymore. The dealers buy the horses and pass them around a few times more through different owners until either the horses find an owner who can deal with their problems or they end up at the slaughterhouse. When I visited several slaughterhouses, every single one of the horses I would have called gorgeous was a behavior problem, kicking and rearing and going nuts. These were beautiful animals going to slaughter because they couldn't be managed.

Xenophon describes gentle techniques for working with horses. People today should follow his advice so that more horses will never have dangerous behavior problems. Xenophon really understood the horse's emotions.

> The one best precept — the golden rule — in dealing with a horse is never to approach him angrily. Anger is so devoid of forethought that it will often drive a man to do things which in a calmer mood he will regret. Thus, when a horse is shy of any object and refuses to approach it, you must teach him that there is nothing to be alarmed at, particularly if he be a plucky animal; or, failing that, touch the formidable object yourself, and then gently lead the horse up to it. The opposite plan of forcing the frightened creature by blows only intensifies its fear, the horse mentally associating the pain he suffers at such a moment with the object of suspicion, which he naturally regards as its cause.[19]

Horse Whisperers Perceive Sensory Details

In the past twenty years horse whisperers and *natural horsemanship* methods have become very popular. The basic idea behind natural horsemanship is to get the horse to accept a bridle, saddle, and rider without using pain or restraint. Monty Roberts, one of the best known of the horse whisperers, says he needs only half an hour to accomplish what other trainers take four to six weeks to do. In half an hour he can get a never-been-ridden horse to calmly accept a saddle and rider on its back. His work with horses became so well known that Queen Elizabeth invited him to demonstrate his methods to her.[20]

I've watched Monty Roberts and some of the other horse whisperers work. They do a really good job training horses.

The interesting thing, though, is that a lot of them are a little Asperger-like or they are dyslexic — not Monty Roberts, but some of the others. Asperger's syndrome is "autism with language"; a person with Asperger's syndrome didn't have a language delay as a child but does have the social deficits and the perceptual differences of an autistic person.

People with Asperger's or dyslexia are often good with animals because their thinking is more sensory-based than word-based. Many times I have told my students that to understand animals you need to get away from language and think in pictures, sounds, touch sensations, smells, and tastes. Everybody who works with horses can learn to be more sensory-based in their relationship with horses. Sit in a quiet place and visualize your horse. What does he look like when he is doing different things? What does his breathing sound like? How does his skin feel when you stroke him? Then when you work with your horse, keep the sensory details in your mind.

An observant trainer sees gradually escalating signs of fear and changes his training procedure before the horse explodes into rearing or bucking. I watched a trainer who was starting a young colt and teaching him to accept a rider on his back. The trainer ignored the colt's tail switching, which got faster and faster until finally the colt bucked the rider off. The colt's fear had built up over a fifteen-minute period like a pot coming to a boil, and the session should have been ended well before the colt reached his boiling point.

In his book *True Unity,* Tom Dorrance, the horseman whom most people see as the originator of natural horsemanship, says, "The older I get the more it's beginning to dawn on me how most people seem to have so little *feel of the whole horse* — of what's going on in what part. Maybe, overall, they have a pretty good idea, but they don't know about this little spot here and that little spot there. Any one of them can be a good little spot, or it can be a bad little spot. People may have an idea what comes out of it, but what's really going on through all of this they don't recognize."[21] A horse whisperer talks about "true unity" or "joining up" or "getting in tune with the horse," but many students do not understand what he means by this.

I have talked to several horse whisperers who are frustrated that their students do not understand their instruction. Horse whisperers see tiny changes in the horses' posture and they can feel the horses' rising fear. Instead of describing to their students the visual changes such as a higher head or auditory changes such as heavier breathing, they describe the horses' emotion. Some students have a natural feel for horses, but others need to be told how to read changes in body posture and breathing. Some trainers do not consciously see the tiny changes; they just feel them. That is why they use vague descriptions when they teach students.

Horse welfare depends on good training. If everyone could train and handle horses the way the horse whisperers and the old-time horsemen do, lots fewer horses would be put down, sold, or neglected because of behavior problems. But that can't happen if people don't understand what the horse whisperers and horsemen are actually doing.

Learning Theory, Negative Reinforcement, and the FEAR System

The real secret of horse whisperers and expert horsemen is that they understand the behaviors associated with different emotional states and they have also figured out that a reward or a cue has to be given within one second after a desired behavior occurs for the horse to make the association. Expert horse trainers understand the horses' emotions, instinctual natural behavior patterns, and the principles of behavioral training. They can do what they do without knowing the science behind it, but everyone else needs to know what the really good horsemen are doing and why it works.

Novice and average-ability horse trainers need to know learning theory because traditional horse trainers often blame the horse for training failures. The really good ones don't, but a lot of trainers do. People have always talked about "willing" horses versus horses that have "vices." Paul McGreevy, author of *Behavior: A Guide for Veterinarians and Equine Students,* says some equestrian books even talk about horses being "depraved."[22] These explanations are given for why a horse is or isn't getting trained successfully. A student has no way of knowing if a depraved horse is frightened, angry, or simply does not understand what the trainer wants him to do.

That's one reason why new trainers should learn about operant conditioning and the power of positive reinforcement. Behavioral trainers never talk about vices and depravity. Behaviorists are some of the most "optimistic" teachers and trainers there are, because if a person or an animal isn't learning, a behaviorist is trained to examine what *he* is doing wrong, not what the person or animal is doing wrong. This means that behavioral teachers and trainers don't blame the student.

The main reason why the behavioral approach is important for horse trainers to learn, though, is that if a horse is failing to learn a new task, learning theory gives the trainer a way to analyze what's going wrong. Learning theory gives people a set of training tools for their horse *and* a set of troubleshooting tools to help them figure out what's going wrong.

Almost all horses have been trained using *negative reinforcement,* not positive reinforcement.[23] Negative reinforcement isn't the same as punishment. The difference is that punishment makes a behavior *less* likely to occur in the future — or it's supposed to — whereas negative reinforcement makes a behavior *more* likely. You can remember the definition of negative reinforcement by thinking about the meaning of *reinforcement,* which is to make something stronger. Troop reinforcements make an army stronger, and when you put steel reinforcing bars inside concrete you make the concrete stronger. Negative reinforcement makes a behavior stronger. It's more likely to occur again in the future.

Negative reinforcement occurs when something "bad" either *stops happening* or *doesn't start happening* as a result of the behavior. I put "bad" in quotes because the bad thing can be very mild, such as the trainer putting pressure on a horse's sides with his legs. The horse's natural tendency is to move away from pressure, but pressure doesn't hurt the horse when the horse first experiences it. Pressure probably arouses the FEAR system to a small degree. When a colt moves forward when a trainer stops pulling on the lead rope, he is negatively reinforced for moving forward. He's more likely to move forward again the next time the trainer pulls on the rope.

Horses are trained primarily through negative reinforcement along with some *secondary positive reinforcers* such as saying, "Good boy!" A secondary reinforcer is a reinforcer the animal has learned to like because it's been associated with something it naturally likes, such as food. Anything an animal naturally likes — food, water, sex, companionship, or being stroked — is a *primary reinforcer.* If you use stroking for a primary reinforcer, you should use firm strokes that feel like the mother's tongue licking her foal. Do not use pats, because horses may interpret pats as hitting.

This is different from standard dog training, which uses a lot of positives along with the negatives. Even rough trainers who do alpha rolls to dominate a dog use lots of food treats. Horse trainers almost never use food treats, mainly for practical reasons. You can't just toss a horse a food treat the way you can a dog, especially if you're sitting on the horse's back. Also, horses are grazing animals that spend seven or eight hours a day eating. They are extremely oral and can get too excited and worked up over a food reinforcer to stay focused. That doesn't happen with a dog.

I think there may be another reason why horses have usually

been trained using negatives, which is that, for a horse, negative reinforcements are stronger than positive reinforcements. Researchers studying horse learning have found that a horse will choose to get away from a negative over staying to get a positive.[24] Horses positively reinforce their trainers for using negatives.

That brings me back to the horse's FEAR system. Negative reinforcement works by activating the FEAR system, and a horse is a big, strong, high-fear animal whose major mode of self-preservation is flight. Horses are built to run away from bad, scary things, and any activation of the FEAR system is going to be highly motivating to them.[25]

The Trouble with Negative Reinforcement

The trouble with negative reinforcement is that many people end up using more and more pressure until it becomes abusive. I have seen this when a colt is trained to lead. When the colt refuses to move, the trainer pulls harder and harder on the lead rope and the colt may get more and more stubborn and scared. I have seen this escalate into pulling an animal with a vehicle. I tell my students over and over that when you train any animal to lead, you give a gentle tug on the lead rope, and if the animal takes one baby step forward, you instantly release the lead rope to reward the animal for stepping forward. Many people make the mistake of continuing to pull.

When a brilliant trainer uses negative reinforcement, there's no problem. Most of the horse whisperers and natural horsemanship trainers (not all) use extremely subtle forms of negative reinforcement that gradually turn into gentle signals. *Advance and retreat,* where the horseman moves just close enough to the horse to cause him to move away in the direction the horseman wants him to go, works through negative reinforcement. When the horse moves, the horseman stops advancing. A negative thing stops happening.

Most people can't use negative reinforcement so subtly. Negative reinforcement in a training situation goes both ways — from the trainer to the horse and from the horse back to the trainer. It's easy to get a vicious cycle of escalating negatives started that way.

This happens in everyday life all the time. A good example is parents and children. A lot of parents feel that they yell at their kids more than they want to, but it's hard to stop. A behaviorist would say that's because parents keep getting negatively reinforced for yelling. Every time a parent yells at a child for doing something bad and the child stops doing whatever he's doing, that is negative reinforcement. The kid's behavior is painful for the parent and yelling makes the painful thing *stop,* which makes yelling more likely to happen in the future because it got results. Yelling has been reinforced by the kid stopping what he's doing. But then, because the parent yells so much, the kid starts to habituate to yelling. He gets used to it. The kid stops responding to being yelled at, so the parent yells louder, and then the kid does respond. That reinforces the parent for yelling louder, and the kid habituates to louder yelling, and so on.

Even when you don't get into a vicious cycle, negative reinforcement often has a downside. The most important drawback is that fear is a very painful emotion for all animals, and you don't want to base your relationship with an animal on fear, especially not with a high-fear prey animal.

Negative reinforcement used the way a lot of people use it — not just animal trainers but parents, teachers, and bosses — has bad side effects. Karen Pryor says negative reinforcement "puts you at risk for all the unpredictable fallout of punishment: avoidance, secrecy, fear, confusion, resistance, passivity, and reduced initiative, as well as spillover associations, in which anything that happens to be around, including the training environment and the trainer, becomes distasteful or disliked, something to be avoided or even fled from."[26]

Using Positive Reinforcement to Turn On SEEKING and Turn Off FEAR

Using positive reinforcement is a much better way to teach or train any animal or person. Positive reinforcement can include treats and stroking. I especially like clicker training for teaching complicated tasks and sequences of movement.[27]

Just on a practical level, the clicker is a help to horse training because it lets the trainer work without food treats for a stretch of time, although you do have to maintain the association between the click and the treat. You can't just charge up the clicker and throw the treats away.[28] But the more important benefits of clicker training are in the way it affects the horse's emotions and the trainer's ability to communicate with the horse.

Clicker training starts when the trainer *charges up the clicker* by pairing the click sound with food. That makes the click into a secondary reinforcer that tells the horse, "Something good is coming." As soon as the click means "something good is coming," the clicker has the power to turn on the SEEKING system, which feels really, really good to all animals. Instead of the horse getting a reward each time he produces a certain behavior, he gets to *anticipate* the reward, which is even better.

Turning on the SEEKING system is a good thing to do when you're training any animal or person, but it may be the most powerful with high-fear prey animals. This means that when you clicker-train a horse you have double protection against his FEAR and flight system. First, when you substitute positive reinforcement for negative, you're not turning on the horse's FEAR system in response to the particular behavior you are teaching. If you are teaching a horse to lift his foot for shoeing, he lifts it not because he is afraid of getting hit by the farrier but because cooperating means something good will happen.

And second, when your horse becomes really accustomed to clicker training or other positive reinforcement to keep the

SEEKING system turned on, you inhibit the FEAR system *overall* because the SEEKING system and the FEAR system are opposed inside the brain. If you're in the middle of a clicker-training session and a piece of plastic blows into your horse's face, he's going to be less likely to panic than he would be if his FEAR system were already mildly "turned on" through negative reinforcement. It's easier for a horse to be brave when he's feeling happy than when he's feeling nervous or afraid. Alexandra Kurland, who wrote *Clicker Training for Your Horse,*[29] says, "Clicker training . . . teaches emotional self-control, and that's huge. If you want a safe horse, clicker training is a great way to build in mannerly, reliable behavior."[30]

This is why equine scientists such as Paul McGreevy believe that scientifically based training techniques could save a lot of horses' lives. A trainer using negative reinforcement who understands learning theory should be able to recognize the signs that he's getting negatively reinforced by the horse to escalate his signals. A trainer using positive reinforcement can avoid activating the FEAR system *and* turn on the SEEKING system in skittish, easily startled horses that might be almost impossible to train otherwise.

A lot of people who use clicker training have talked about how much happier and more enthusiastic their animals are when they use positive reinforcement for training. That's because any form of positive reinforcement training uses *shaping.* Shaping is when you reinforce behavior the animal is already doing naturally, on its own. If you're training a lab rat to press a bar, for example, first you reinforce the rat for turning its head ever so slightly in the direction of the bar. Then, when the rat has learned to turn its head slightly toward the bar, you reinforce it for turning its head farther in that direction. You keep on going until the rat is pressing the bar.

Negative reinforcement is the opposite. Using negative reinforcement, the trainer pushes or pressures the animal into the behavior he wants and *then* reinforces the behavior by relieving the pressure, giving the animal a food treat, or both. You can use negative and positive reinforcement together, and a lot of people do. With positive reinforcement, the animal suddenly "gets it" — realizes that it can do something to make a good thing happen. That's called *learning to learn.* When the animal learns to learn, it starts to *offer behavior.* That's what behaviorists call it. It'll intentionally run through all kinds of different behaviors looking for one that will work.

Karen Pryor says animals that have learned to learn start to feel like they're training the person, not vice versa. They know they can figure out a way to make the trainer give them treats. She could be right, based on what behaviorists know about humans. I read a very interesting article by three research psychologists on *positive* versus *aversive* control of people.[31] *Aversive control* is what they usually have in a public school. The students have to do their work and behave well in class or they'll get bad grades or a detention. *Positive control* might be used in a preschool, where the teachers "catch them being good" and then reinforce the good behavior. Instead of making the children do good behaviors by threatening to punish them if they don't, the teachers watch the children until they spontaneously do a good thing and give them rewards to reinforce the behavior and make them more likely to do that behavior again in the future.

The psychologists said that people feel different under these two systems. When a person is under aversive control, he *feels* like he's being controlled. The authors write, "The person reports that his or her autonomy was undermined because avoidance or escape behaviors are verbally understood as things that he or she 'had' to do."[32] Positive control is the opposite. Even though the teacher or psychologist has created an environment that "controls" the person's behavior through positive reinforcement, the person doesn't feel like he's being controlled, probably because he is getting reinforced for behaviors he didn't "have" to do. The authors say: "The behavior is likely to be reported as having been the product of an autonomous decision to act. Subjectively, behaviors that are followed by pleasing consequences are likely to be verbally described as those that we 'like' to or 'chose' to engage in."[33]

Karen Pryor has a story about using positive reinforcement to retrain an Arab horse who was refusing to prick up her ears in the show ring that shows how an animal starts to offer behavior on purpose:

> Unfortunately there are still some trainers who swish a whip around the horse's head . . . [*negative reinforcement*] to get it to prick up its ears. This mare, however, had gone past that. When the whip was swished she'd taken to pinning her ears . . . and baring her teeth . . . And of course escalating the swishing just escalated the ugly face.
>
> The new trainer had begun clicking the mare for pricked ears . . . and the mare had learned that clicks mean carrots; and that she could make clicks happen. And she had also become aware that the behavior the trainer wanted had something to do with ears. But what? So she's doing this: flopping her ears this way and that, backwards and forwards, one ear up and one ear down, rotating each individually and then together — quite a show.[34]

Animals trained using positive reinforcement learn faster, too. If you put a horse in a maze and let him find his way out through trial and error, he'll finish faster than a horse who gets a shock when he makes a wrong turn. Paul McGreevy says, "Punishment can stifle creativity and impede a horse's innate problem-solving skills."[35]

Retraining Fearful Horses Using Positive Reinforcement

So far, one study has been published that supports the idea that clicker training and positive reinforcement inhibit the FEAR system. This was a study where the researchers used clicker training to retrain five horses that were problem loaders. A problem loader is a horse that refuses to get into a horse trailer. This is a common and very dangerous problem in horses, probably because horse trailers are intrinsically frightening to them. Earlier, I talked about the importance of making a horse's first trailer experience a positive one. A horse trailer is a very tight, confined space, and horses in the wild avoid confined spaces they can't easily escape from. A trailer is usually dark, too, which makes it scarier. Most horses wouldn't naturally walk inside a horse trailer, and when you add in the fact that handlers use aversives to train horses to load, you end up with a lot of problem loaders. If a horse gets violent enough that his owners give up and don't load him, the horse has also been negatively reinforced for refusing to load.

None of the horses in the study had had traumatic experiences while loading onto or riding in a trailer, such as falling down or being involved in a car crash, and all five were pedigree quarter horse mares. Quarter horses usually have temperaments that are in between the hot-blood, high-fear Arab-type horse and the cold-blood, low-fear draft horse. The researchers in the study were trying to remediate fear, not terror or panic. They weren't working with the kind of horse I call a "fear monster."

These horses were "fear protesters." They were on strike. The authors say, "The horses were selected because their loading required a significant amount of time (up to three hr) and effort . . . All five horses had been forced into trailers in the past, through the use of whips and ropes."[36] The goal of the study was to get all of them to load themselves on a verbal command from the owner, and all five horses reached the goal. That's amazing and very unusual. The trainers totally changed these horses' entrenched FEAR behavior using positive reinforcement *exclusively.*

While they were running the study, they also noticed that the horses' behavior and attitude improved in other areas outside of the loading situation. Two of the horses had always run to the back of the pasture whenever they saw anyone holding a halter and lead rope. Then they wouldn't hold their heads still for the halter to be put on. Not too long after the clicker training started, both horses began to come up to the fence instead of running away when they saw the halter and rope, and they even began to put their heads down to make getting the halter on them easier.[37]

I want to give one word of warning. In many cases, a horse with a severe fear memory cannot be completely fixed. The SEEKING system can inhibit the FEAR system, but fear memories have a nasty habit of coming back. For a behavior such as trailer loading, a horse that successfully loads 95 percent of the time after retraining is a great success. For more dangerous behaviors such as striking with the front feet, a horse that is good 95 percent of the time can still land a person in the hospital during the one time out of twenty when he strikes. Careful habituation and clicker training should be used to *prevent* dangerous behavior problems from developing.

Clear Communication

The other benefit of clicker training is that the clicker lets most people communicate with horses much more clearly and precisely than they've been able to in the past. Alexandra Kurland says that when she first started using clicker training with her horse, "I could almost feel him saying, 'Oh, that's what you wanted me to do! Why didn't you say so before?' His training took a huge jump forward. All the little glitches and misunderstandings melted away . . ."[38]

I think one of the reasons clicker training can be so effective is that it tunes in to the animal's hyper-specific sensory-based thinking. The horse can associate the click with a specific visual, auditory, or touch sensation he is experiencing the moment he hears the click. Teaching a horse to put his ears forward is a simple way to learn how to use clicker training. When the show horse that Alexandra Kurland knew was being trained to put her ears forward, a click was given the instant the ears started to move forward. A basic principle of clicker training is to give a click when the animal makes a first response that is only part of the desired behavior. The horse quickly learned that pinned ears never received a click, so she started to experiment with different ear positions to SEEK a click. To the horse's sensory-based hyper-specific thinking, having the ears pricked and pointed forward was different from having one ear forward and the other rotated sideways. When both ears came forward she got another click. The horse quickly learned that *both ears forward* was always rewarded with a click. You could even teach your horse to point one ear toward a flag. This would be kind of silly, but you would really learn how to use the clicker to teach a behavior that is more complex.

True Communication with the Horses

A few years ago I saw Pat Parelli, one of the natural horsemanship trainers on the circuit, give a demonstration at the Rocky Mountain Horse Expo in Denver. He rode a beautiful sleek black horse bareback without a bridle or a halter. The horse went through all the gaits — walk, trot, and canter — and turned right and left at Pat Parelli's signal.

If he wanted the horse to go faster, he leaned forward. If he wanted the horse to go slower, he leaned backward a little. If he wanted the horse to stop, he leaned back a little more. When he wanted to turn right, he turned his head to the right; if he wanted to turn left, he turned his head to the left. When Pat turned his head, the horse felt a faint directional signal from his legs. This was a beautiful, calm horse. No tail flicking, no skin quivering — totally peaceful. That's what positive horsemanship looks like. I hope someday all horses can have such a beautiful relationship with a person.

5 Cows

DOMESTIC CATTLE AREN'T AS HIGH-FEAR as horses, but they are ever vigilant for predators. They have wide-angle panoramic vision, and a cow's brain is like a sentry that instantly sees small, rapid movements that may signal danger. The big difference between cattle and horses is that cattle aren't pure flight animals. When cattle are threatened by a predator, they bunch together and seek safety in numbers or turn and fight with their horns. That may be one reason why cows aren't as high-fear as horses.

Cows are herd animals that need to be with their buddies and family members. They have close relationships, especially between sisters and between mothers and daughters, who like to graze together. In today's herds most of the bulls have been removed to be raised for meat so their natural social behavior cannot be observed. Cattle stay together in groups to make their daily rounds between the pasture and the water trough. They walk leisurely along from place to place in single file. The cow leading the line *isn't* the dominant cow. That's a mistake most people make. They assume that the cow that is the boss at the water trough will also be the leader of the herd when it moves from one place to another. But the "leader cow" is usually a curious, bold cow whereas the dominant animal, which pushes the other cows away from the water trough, stays safe from predators by walking in the middle of the line.

The leader cow isn't really a leader at all. The reason she is at the front of the line is probably just that she has high SEEKING emotions and low FEAR. She is not telling the group where to go. Herd animals seem to make decisions about movement democratically. Red deer move when 62 percent of the herd has stood up, not when one "leader deer" has gotten up and signaled everyone else to move. One study of resting and activity switch-offs in a herd of beef cattle found that "there were no leading animals initiating switches in activity in our herd." Instead, their behavior was "highly synchronized," with cattle of the same body weight being more synchronized in resting versus moving than cattle of different body weights. No one understands how this works.[1]

The "dominant" cows aren't really dominant, either — not the way most people think of dominance. The dominant cow is the one that gets the food when there's pushing and shoving at the feed trough. The subordinate cow is the one that doesn't get the food or that has to wait and get the food later. All cattle herds have dominant and subordinate cows that you can pick out by watching them butt each other and push each other away from feed. But it looks like dominance is a *social* characteristic of the relationship between animals, not a *personality* characteristic of an individual animal.

A group of researchers did a really interesting study of social dominance and feeding-related displacements, which means one cow bumping another cow out of its place at the feeder trough. Usually you expect social dominance to be asymmetrical and transitive. *Asymmetrical dominance* means that dominance goes in one direction. Cow A dominates Cow B, and Cow B never dominates Cow A. *Transitive dominance* means that if Cow A dominates Cow B and Cow B dominates Cow C, then Cow A also dominates Cow C.

But in this study 52 percent of cow interactions at the feed trough were bidirectional. Sometimes Cow A displaced Cow B from the feed trough; other times Cow B displaced Cow A.

Forty-five percent of all the cow triads were intransitive: Cow A dominated Cow B, Cow B dominated Cow C, and Cow C turned around and dominated Cow A.[2] That's not what you'd expect at all. So we know that cows are social animals, but we don't know very much about the social structure of the herds they live in.

Mama cows are very protective of their calves, and a mama cow will turn and fight a coyote on the range. I visited a feedlot in Brazil where the mamas were penned with their calves. The calves were small enough to squeeze under the fences, so when no people were around they'd crawl out and visit the cows in the other pens and look for grass to eat. But the minute visitors came, the moms would start bellowing and the calves would come running home to mama and scooch back under the fence into the pen. It was as if the mama cows were shouting at their little kids, "*You come back here right now!*"

The FEAR System in Cattle

A lot of times people who haven't grown up around farm animals don't realize that most beef cattle aren't tame. Pet cows and working oxen *are* tame, and dairy cows that are milked two or three times a day are somewhat tame because of both close association with people and lower fear genetics. The Holstein dairy cow is probably genetically further away from her wild ancestors compared to beef cattle because she has been selected solely for milk production. But most beef cattle on the big ranches are not tame. Cattle only seem tame because they don't take flight the instant they see a person. But what they really are is *habituated* to the sight of human beings, not tame. There were 97 million head of cattle living in the United States on January 1, 2007;[3] there would be no way to completely socialize every one of 97 million head of cattle even if you wanted to.

If you get close enough to a cow, it will run away. How close a wild animal lets you come before it moves away is called the *flight zone*, and the more fearful the animal, the larger its flight zone is. Tame animals have flight zones of zero. That's one of the behavioral definitions of tameness in an animal: a flight zone of zero for humans. Cattle, because they live around humans and are used to seeing them, have a smaller flight zone than they would if they were living in the wild where they would see no humans at all.

The fact that many farm animals aren't tame means that FEAR is usually a more important welfare issue for them than it is for companion animals, because farm animals are always a little nervous around humans no matter how nice and unthreatening people act.

Even if cattle weren't innately fearful of humans, the rough handling a lot of cows experience would put FEAR on the top of the list of bad emotions cattle experience too much of the time. A central welfare issue for beef cattle is poor stockmanship. People screaming and yelling at cattle, hitting and punching them, shocking them with electric prods — all of these things terrify cattle. Some livestock handlers treat cattle so badly that I had to include five automatic "fails" in the animal welfare audit I wrote for the American Meat Institute in 1997:

- hitting or beating animals up
- dragging live animals
- driving animals on top of each other on purpose
- sticking prods and other objects into sensitive parts of animals
- slamming gates on animals on purpose

None of these things should ever be done to any animal.

Understanding What Scares a Cow

Anyone who believes that animals have feelings would expect actions like dragging an animal with chains to be terrifying. But people have a hard time understanding that actions that seem only mildly negative or even neutral to human beings can be very frightening to cattle. Paul Hemsworth, an Australian scientist who has researched fear in cattle, says, "The negative impact of interactions that are routinely but briefly used by stockpeople, many of which intuitively appear to be innocuous and innocent, is . . . surprising. Even after 20 years of research on this topic of human-animal interactions, I am still surprised at times . . ."[4]

Paul Hemsworth lists five moderate stimuli that frighten cattle:

- yelling at the animal
- sudden appearance of a human in the animal's field of vision
- a human "looming over" an animal
- fast movements (cars, bikes, predator animals like wolves)
- sudden movements (a branch falling from a tree; any unexpected movement whether it's fast or slow)

The only item on this list that's obviously negative is yelling. Cattle *hate* being yelled at. What's frightening isn't the noise so much as the person's anger. In one study, cattle's heart rates and restless movements were greater when the cattle were listening to a recording of people yelling versus when they were listening to an equally loud recording of metal clanging.[5] Cows *know* when people are mad and it scares them. Compared to dogs, cattle and sheep are probably more afraid of angry people because of the low-level fear they feel toward human beings. It's like the difference between the timid, shy child in a family and the rowdy, hyperactive child. If the mom yells at the kids, the shy child might start crying while the rowdy child barely even hears her. Cattle being untamed, they're more like the timid, shy child biologically when they're in the presence of humans. Another factor is that prey species animals such as cattle and horses are genetically more fearful than most dogs. Since their nervous systems are tuned to detect potential threats, they startle more easily.

The other four items on the list — sudden appearance of a human in the animal's field of vision, a human standing over an animal, fast movement, and sudden movement — sound harmless, but they aren't harmless to a cow that's already on edge around humans. Cattle are like deer that live in the suburbs and have habituated to people. A suburban deer doesn't take off running the second it sees you, but if you happen to suddenly pop up in its field of vision, it would get scared and bolt. That's fine for a deer in the field, but it's dangerous for a 1,500-pound cow being handled on a ranch.

The difference between a flighty temperament (high-fear and high-reactive) and a calm temperament (low-fear) is important in both horses and cattle. Cattle with calm genetics such as Holstein dairy cows or Herefords usually have a smaller flight zone than the more flighty breeds such as Salers. A tame, flighty cow may go berserk the first time it sees a speeding bike, and the animal with calmer genetics may just flinch.

Understanding How *Universal* Fears Apply to Cattle

Paul Hemsworth also cites a useful categorization for types of fears shared by all animals, including people, which was developed by Dr. Jeffrey Gray in the 1980s. According to Dr. Gray, who is at the Institute of Psychology at King's College in London, fears fall into one of five categories:

- high-intensity stimuli
- special "evolutionary dangers"
- socially learned fears
- learned fears that are acquired when a neutral person, thing, or situation is associated with something bad
- novel stimuli

People working around cattle can benefit from keeping this list in mind and from recognizing how it applies to cattle specifically.

High-intensity stimuli: Yelling is frightening partly because it's a socially learned fear and partly because it's a high-intensity stimulus. Cattle like quiet people and quiet handling.

I have observed that high-contrast images such as shadows or a gray fence with a bright yellow raincoat flung over it are other examples of high-intensity stimuli that spook cattle when they are forced to move toward them. Cows have dichromatic (two-color) vision, and yellow is the highest-contrast color for them, so yellow things — raincoats, warning signs — scare cattle.

Evolutionary dangers: People and animals are naturally afraid of heights, isolation, snakes, and so on. Not all animals have the same evolutionary fears. Small prey animals such as mice feel safe in small, dark places, for instance, whereas large prey animals that rely on flight, like horses, naturally avoid strange, closed-in, dark places because they may fear being trapped.

Cattle are naturally afraid of heights and rapid movement. Objects that move rapidly can instantly turn on the FEAR system in cattle and other prey species animals, but rapid movement turns on the SEEKING system in cats and dogs when they chase after a ball or a furry little animal.

Socially and individually learned fears: Cattle learn to fear people, places, and things that have hurt them, and they learn fears from each other.

Novel Stimuli: Why Animals Are Curiously Afraid

The fifth category, novel stimuli, isn't very well understood. *All* brand-new, never-before-seen things in the environment are potentially frightening to *all* animals. Cattle react to novel stimuli by acting *curiously afraid.* A friend of mine invented that expression. She was with me when we saw a cow very cautiously investigating a yellow raincoat slung over a fence and my friend said, "It's like she's curiously afraid." I've been calling it that ever since. Cows want to investigate anything new or out of the ordinary, but they're also afraid it might be dangerous or bad.[6] One time I put a clipboard with a sheet of paper on the ground near a group of cattle. They really showed me their curiously afraid behavior. The boldest animals touched the paper, but when the wind made the paper flap, they instantly backed away.

I don't know how this works inside the brain. It might be that their brains were switching back and forth between FEAR and SEEKING. The novel stationary object turned on SEEKING and the sudden movement turned on the FEAR system. The cows acted like their brains had a switch because every time the wind blew, they jumped back, and then, when the paper stopped moving, they cautiously approached.

The problem with that explanation is that SEEKING tends to inhibit FEAR, and a cow that is acting curiously afraid looks like it is feeling some of both emotions at the same time. I don't think it is feeling the two emotions at the same time, however. Instead, its brain may be quickly switching back and forth between SEEKING and FEAR.

There is some research out on the amygdala, which is the fear center in the brain, that might help explain why novelty is both scary and attractive when you put it together with what we already know. Starting with what we know, researchers have done many experiments showing that novelty and unpredictability are attractive to animals and people. Peter Milner, a psychologist who studies attention and learning, says that the human brain is programmed to be constantly finding out whether unpredictable things are good or bad, which means that the brain is programmed to pay attention to and explore unpredictable things. In his book *The Autonomous Brain*, Dr. Milner points out that animals and people constantly choose to do things whose outcome cannot be predicted.[7] Monkeys will work for a chance to watch a toy train going around a track or even just to be able to see what people are doing in the next room. According to Dr. Milner, the brain's default setting is: If nothing is investigated, nothing is gained. If animals or people can't predict whether an action will have a good result or a bad one, they go ahead and perform the action. New things are always unpredictable, so I conclude that animals and people are programmed to pay attention to and explore new things.

The study on the amygdala shows that the brain's FEAR system is turned on by *uncertainty*, as well as fear,[8] and that the emotion aroused by uncertainty is probably a mild anxiety or *vigilance*. That describes curiously afraid cattle. When they see something new, they act vigilant. They want to explore the new thing and find out what it is, but they're also scared that it might be a bad thing. They're curious and a little bit afraid at the same time.

In the first part of the new study, the experimenters played tones at either predictable or unpredictable intervals for mice and people. Both mice and people showed greater activation of the amygdala for the unpredictable tones. Unpredictable tones turned on the amygdala.

In the second part of the study, the experimenters put the mice and people in standard anxiety-testing research paradigms and played one of the two sets of tones in the background. Naturally anxious mice and humans always behave differently in these tests from naturally calm mice and humans.

The mice and humans in the new study weren't naturally anxious, and when they were put in the tests with the predictable sounds playing in the background they didn't act anxious. But when they heard the unpredictable sounds playing in the background, suddenly they acted the way naturally anxious mice and people act when put in these research paradigms. Unpredictable sounds turned on the amygdala, and the mice and people acted the way naturally anxious mice and people do in these paradigms.

The amygdala doesn't just respond to dangerous things that

are definitely scary. It also responds to unpredictable things by making a person or animal feel slightly anxious or vigilant. Paul Whalen, a professor at Dartmouth who studies fear learning and the amygdala and who wrote a review article about the study, has a good way of thinking about why the fear center would be activated for novel stimuli:

> Fear is important. Without that rush of fear you felt the first time the subway train roared by, you might not remember crucial things, such as standing behind the yellow line rather than in front of it. But as important as fear is, . . . you can probably count the number of times in your life you have been "scared out of your wits" on your fingers and toes. So what do brain areas important for fear-processing "do" the rest of the time?
>
> . . . What role exactly does the amygdala have in the complex chain of events that includes monitoring the environment, detecting a stimulus, registering it as threatening, feeling afraid and devising a plan of action?[9]

Dr. Whalen concludes that the new study shows that "at least a portion of the healthy amygdala acts as if it has an anxiety disorder — searching for threat in response to uncertainty."[10]

Forced Novelty Is Frightening

My interpretation of the behavior of cattle is there are two parts of the FEAR system. One produces anxiety or vigilance, and the other is the full fear response. When the cattle acted curiously afraid, their vigilance system was activated but the full fear response had not been turned on. When the paper on the clipboard moved, the cattle may have been switching back and forth between SEEKING and vigilance. As a person with autism, I can really relate to this through my experiences with antidepressants. The medication stops my panic attacks and blocks my fear response, but my vigilance/anxiety circuits are still hyperactive.

When I hear a strange noise at night, I am instantly awake, but the medication has stopped the heart-pounding fear.

When you're working with animals, novelty can be attractive or scary depending on how it is presented.[11] The single most important factor determining whether a new thing is more interesting than scary is whether the animal has control over whether to approach the object. Animals are terrified by forced novelty. They don't want new things shoved into their faces, and people don't either. But if you give animals and people a new thing and let them voluntarily decide how to explore it, they will.

Unfortunately, in a meatpacking plant cattle don't have the freedom to slowly investigate scary new things. They have to keep moving. So you want to have *no* novel stimuli inside a meatpacking plant. In *Animals in Translation,* I have a checklist of items that must be removed from the handling chutes to avoid triggering the FEAR system. Some of them are reflections on shiny metal, coats on fences, flapping paper towels, and dangling chains. Cattle will move easily through the plant when these novel distractions are removed.[12] On a pasture or in a feed pen, a novel item such as a water bottle can be more interesting than scary because the cattle can control how close they come to it and how thoroughly they explore it.

The FEAR System and Moving Cattle from One Place to Another

The two times cattle are most likely to be upset by their human handlers are when they have to be moved and when they have to be closely inspected and restrained for a veterinary procedure.[13]

Moving tame cattle isn't a problem. You just carry a bucket of grain in your hand and they follow you. Getting cattle to enter restraining devices or trucks isn't hard either, if the cattle are carefully trained with positive reinforcements such as yummy feed treats. I have trained both cattle and sheep[14] to line up at a gate and wait their turn to enter the squeeze chute. The sheep jumped up on the gate because they wanted the grain. Ranchers who frequently use trucks to take cattle to new pastures find that loading is easy. The cattle know the truck will take them to new delicious food. When they see it, they come running. In all of these cases the SEEKING system has completely replaced FEAR as the reason why cattle cooperate with their handlers.

Herding cattle over a distance is a harder problem. Cattle grazing *intensively managed pastures* are rotated so frequently among pastures that they can be led from one pasture to another. The ranchers don't have to drive them. Unfortunately it is simply not practical for all beef cattle living on miles and miles of rangeland to be handled to the point where they become tame and actually SEEK handling procedures.

To drive wild cattle over a long distance, you have to activate the FEAR system to the smallest possible degree, and that's hard to do quietly. The old roundup scenes in movie Westerns, with horses running back and forth and cowboys whistling and yelling and lots of mooing, show the absolute wrong way to move cattle. Those are very stressed cattle and in real life their fearful behavior would stand a very good chance of escalating. Anytime cattle start to run, it's bad, especially if there are mamas and babies in the herd, because the calves can't run fast enough to keep up. Being separated from mama is a gigantic stress for baby, who may gain less weight because of fear and stress.

Once cattle have broken into a run, they can progress to *milling*, which is a very dangerous situation. Milling results when the outside cows try to get to the middle of the herd because the wolves are biting around the perimeter. The strongest animals try to shove into the center, and the whole herd goes around and around in a circle and totally jams together. Milling cattle will trample anything in their path. Milling is a herding animal's ultimate defense when a predator is attacking and the FEAR system is in overdrive.

Poor practices when moving cattle on the range are a major welfare issue for cattle. Steve Cote, a range conservationist with the Natural Resources Conservation Service, talks about the problems in his book *Stockmanship: A Powerful Tool for Grazing Land Management*:

> I saw a stampede one July day when 1,500 cows and calves were driven 12 miles. Then they ran back 12 miles because they ultimately couldn't take the pressure of being driven through a gate . . . The riders couldn't get ahead of the dust to stop the stampede. They handled the drive and going through the gate the way most people handle cattle around here.
>
> I'm sure these cows didn't want to stampede in the middle of a hot summer afternoon, and I'm real sure the riders didn't want them to. The riders made a gather and tried the same drive again the next day . . .
>
> They had a few extra riders and held the cattle from breaking back. But . . . at least 125 calves had to be doctored for pneumonia the following week.[15]

The Cow Whisperer

To herd untamed cattle correctly you "pressure" the flight zone by coming right up to the edge of the zone and slightly entering it. That gets the cattle moving away from you. Then you drop back when the cattle are moving in the desired direction. It's the same *pressure-release* principle horse trainers use to train horses. It works by activating the FEAR system. That's why it's so easy for cattle to panic when they're being driven across long distances if people put too much pressure on the flight zone.

In cattle handling, Bud Williams is the cow whisperer counterpart to horse whisperers like Pat Parelli, Ray Hunt, and Tom Dorrance. Bud has been doing workshops on low-stress livestock handling for stockpeople across the country since 1989. Unfortunately, many ranchers have had difficulty understanding his methods because often he does not provide clear directions for how to use them. I have seen many people who are masters at working with animals have difficulty explaining or clearly writing about their methods.

I've been very interested in Williams's methods and have attended several of his workshops. After one of them, which I went to with my student Jennifer Lanier, we sat in the Chinese restaurant at the airport for four hours trying to break the material down into directions other people could understand.

After we finished, I took our directions to my assistant, Mark, to see whether he could herd cattle the way Bud did based on our instructions. Mark and I went out to a ranch with about seventy-five cattle in a flat valley with hills on one side. Mark started out doing the zigzag, back-and-forth movement Bud teaches, walking silently at a steady pace the way Bud does.

The cattle scattered all over creation. This has happened to a lot of ranchers who tried Bud's methods. When Bud zigzagged back and forth behind a herd of cattle, it worked like magic, and the cattle moved forward together as a herd, but when other people tried it, the cattle just spread out across the land.

After the cattle scattered I remembered Bud saying, "Take your time." So I told Mark, "Take your time; don't push them. Just walk back and forth."

Mark walked back and forth without pushing the cattle at all. He stayed barely on the edge of their flight zone. Then the magic happened and they came together as if a magnet were in the middle of the field attracting them. That's how I always describe it when I'm explaining Bud's technique to people: "the cattle magnet in the middle of the field." Once the cows were together in a soft bunch, they could be quietly moved as a herd without any screaming or yelling. They didn't scatter.

The minute I saw the cattle bunch together I said, "This is hard-wired." It was instinctive. I call those kinds of automatic behaviors "nature's little pieces of software." That's when I figured out why people were having so much trouble trying to adopt Bud's approach. Nobody realized you have to trigger the soft-bunching instinct in the cattle before you put additional pressure on the flight zone to move them.

With Bud's method, the cattle aren't Super Glued together in frenzied milling. They're just loosely arranged in a herd formation that gives them more protection while they continue to graze. This is a common behavior in grazing animals that live in big, wide-open spaces where predators can be plainly visible in the distance. If you watch wildlife shows on Animal Planet, you'll see antelopes calmly eating while a pride of lions rests a mile or two away. Prey animals know whether a predator animal is stalking them or not. Cattle that live in country with lots of predators naturally graze in soft bunches. In areas without predators, like a modern ranch, the animals spread out.

Bud Williams doesn't agree with me on this, but I think gentle herding uses cattle's hard-wired *prey-defense behaviors* to move them quietly from Point A to Point B. When a stockperson zigzags back and forth on a straight line just outside the herd's flight zone, he is acting enough like a predator to *mildly* stimulate the anxiety/vigilance part of the cow's FEAR system and turn on the software. That's why the cattle start to form a soft bunch. After the soft bunch forms, the stockperson penetrates the flight zone more deeply to get the herd moving forward. They are treating the person like a predator and gathering together for protection, then moving away as the "predator" moves toward them.

The proof of this is that Bud's methods don't work with completely tame cattle. When I walked back and forth behind a herd of completely tame cattle, they just looked at me like I was stupid. *Tame cattle can't be herded.* They can be led, but they can't be herded, because there is no fear.

Bud has no formal training in animal behavior, but he is absolutely amazing when he moves cattle. He loves cattle, and I think the idea of acting like a predator really upsets him. What makes Bud Williams's techniques so gentle is that he is extremely sensitive to cattle, and once the cattle are moving, he instantly removes pressure on their flight zone. Also, he does everything on the cow's time, not on the human's time. A stockperson trained by Bud Williams will keep quietly zigzagging back and forth for as long as it takes to get the cattle moving. Although a stockperson using Bud's methods does activate the FEAR system to herd untamed cattle, we're talking about extremely low-level anxiety on the order of what an out-of-towner might feel driving a car in an unfamiliar city. You could say the cattle feel a little nervous or on edge, nothing more than that.

After cattle have been herded gently a few times, they become *trained* to go where the stockperson wants them to go. If cattle have been gently herded to places they *like*, they may switch over from the FEAR system as the motivator of their behavior to SEEKING, and now the stockperson has switched from negative reinforcement to positive. The cattle see the stockperson zigzagging just outside their flight zone and they start to move because a zigzagging stockperson means something good is going to happen. They're going to a new pasture with fresh grass to eat. Steve Cote says, "When we use proper handling, the stock get calmer and calmer, and then real control comes. It just happens. The cattle are able and willing to do what we want."[16] Those are trained cattle.

Bud works with cattle so gently that they start to become tame. Their flight zone gets smaller and smaller as they see that nothing bad happens to them during herding.

Riparian Loafers

One of the biggest headaches for cattle ranchers who graze their cattle on government land is riparian loafing. *Riparian loafing* means cattle that keep grazing protected land beside rivers and creeks after they've been repeatedly herded off. The government strictly regulates grazing in riparian areas because healthy creek beds and riverbeds are essential to a healthy range, and agencies will pull the permits of ranchers who can't keep their cattle away from the banks. That puts a lot of pressure on the ranchers because as few as twenty to forty cattle can "grub out" a creek bottom. Steve Cote says riparian loafers are so hard to deal with that stockpeople overwork their horses constantly chasing the cattle back to where they're supposed to be. Some riders use six or eight saddle horses apiece. Even so, these horses commonly lose hundreds of pounds or just give out. Some die.[17]

The reason all these things happen is that riders are working against the cattle's nature instead of with it. Cows always want to go back to a place where they have felt safe and comfortable. That is a core trait in cows (and also in horses). The reason riparian loafers go back to the riverside over and over again after being chased off by a stockperson is that that is the one place where they *weren't* being chased. They will keep going back even when there's almost nothing left there to eat, because they feel safe there.

To prevent cattle from becoming riparian loafers in the first place, or to cure cows of a riparian loafing habit they have developed because of poor handling, riders can't just chase the cattle to where they want them. They have to use pressure *and* release. The rider pressures the cows' flight zone to get them to move away from the riverside, then, *the instant the cows start to move,* he releases the pressure by moving back out of the flight zone. That way the in-between land the cattle are moving through doesn't get associated with being chased. If the rider herds twenty loafers back to the main herd by pressuring their flight zone all the way to the rest of the herd, those cattle will go right back to the riverside the minute he leaves.

You might think that after cows get chased off a restricted area enough times, they'd start to associate that place with being chased and yelled at and not want to go back. But that doesn't happen, very likely because of poor use of the principles of positive and negative reinforcement. When the stockperson chases the cattle all the way back to the pasture where he wants them to be, he's using punishment instead of reinforcement, and he is punishing the behavior of moving in the direction he wants the cattle to move. The stockperson has to use pressure *and* release. The release rewards the cattle for moving where the stockperson wants them to move.

A double experiment with training cattle to stay away from a full feed trough shows how important it is to use reinforcement precisely and with proper timing.[18] In both experiments, researchers put electronic gun-dog collars on twelve cows and tried to use shocks to train them. In the first experiment, the researchers used a pressure-release approach. A cow was given a shock ("pressure") as soon as it crossed over the invisible boundary into the trough area. If the cow backed up, turned around, or stopped, the shock went off ("release"). If the cow kept going toward the feed trough, the shock stopped after five seconds.

In this condition, all of the cattle learned to stay away from the feed trough. On the fifteenth day after training began, one cow reached the trough; on the twenty-ninth day two cows reached the trough. After that all of the cows stayed away. That is classic *associative learning*; the cows had learned to associate the shock with the behavior of going toward the feed trough. When they stopped moving toward the trough, the shock went off and they were negatively reinforced for staying away from the trough.

In the second experiment, the researchers used an *uncoupled* training approach. In this condition, the cow's *behavior* wasn't tightly linked, or *coupled,* to the shock. Instead, the shock was linked with the *place.* Anytime the cows were inside the trough area, they were shocked for five seconds. When the cows were outside the trough, they weren't shocked. There was *some* connection between behavior and the shock going off because the shock also terminated if the cow happened to leave the area, but there was no connection to "intermediate" good behaviors such as stopping, turning, or backing up. The cattle were in an on-off condition. Inside the zone, five-second shock. Outside the zone, no shock.

These cattle didn't learn well at all, and one other study found the same thing.[19] You can't train cattle to stay out of an area they want to go into just by having something bad happen when they're inside the area. The training and communication have to be more precise so the cattle know which specific behaviors they can do to avoid the bad thing happening.

The cattle in the uncoupled training condition were in the same situation as poorly managed riparian loafers. Loafers are happy and content grubbing out the creek bottom when all of a sudden, out of the blue, they get a punishment: a frustrated stockperson riding straight at them and yelling. But they don't make the connection between *grazing in riparian areas* and *angry, yelling stockperson.* With poor handling, the next thing that happens is that the loafers start to walk or run away, and the rider keeps on coming after them and shouting, which means the cows are being punished for doing the behavior the rider wants. This is bad training for any animal or person, but it's especially ineffective with cattle because of their instinct to go back to the safe place.

The sad thing is that cattle handling may have been better over a hundred years ago. The cowboys who drove the great cattle drives across miles and miles of the American West recognized the importance of quiet movement. On those long drives, stress from rough handling would have killed hundreds of cattle. The drives were many times longer than drives today. Ineffective, high-stress cattle handling may be a consequence of people getting too far away from nature and from the knowledge of animals and the land their grandparents had. I'm going to talk about the consequences of lost knowledge later on.

Stockpeople need to relearn what cattle are like and what they can do when they're handled quietly. They will stay together as a herd when they are moved, and the mother cows and their calves will seldom get separated.

Frightening and stressing cattle is bad because it's wrong to treat animals badly, but it's also bad business. People have known this for a long time. Going all the way back to 1925, W. D. Hoard, the founder of *Hoard's Dairyman* magazine, recognized the importance of good stockmanship to milk production.[20] There have been so many studies showing that good stockmanship improves milk production, weight gain, and reproduction. Paul Hemsworth and Grahame Coleman published a whole book on the subject in 1998.[21]

I would like to see all ranchers adopt low-stress cattle-handling methods. The rewards are so large that I think more and more ranchers will adopt them as they see the good results they produce.

FEAR and RAGE

The other high-risk time for cattle welfare is when cows have to be closely inspected. By definition, anytime a human gets close enough to a cow to give it a shot or provide veterinary care, that human has violated the cow's flight zone. Close inspections activate an animal's FEAR system in two other ways: The personnel and the equipment are often novel, and the procedures themselves, such as shots, can be uncomfortable.

In most cases, close inspection also requires restraining the cow, which activates the RAGE system to some degree. The cow's FEAR system can easily escalate the cow's mild frustration from restraint into furious RAGE because the FEAR system has an excitatory effect on the RAGE system.[22] I have seen wild cattle stand quietly in a restraining chute until people came up and stood next to them. That is terrifying to the cow, because it cannot move away as the person invades its flight zone. The first sign that the animal is really upset is when its tail starts switching back and forth. Then all of a sudden violent struggling begins. I think struggling occurs when the cow's intense FEAR at having a person come so far inside its flight zone escalates frustration into fury.

I have worked with many ranches and feedlots on proper handling of the cattle as they are brought up to the squeeze chute. Cattle emotions and behavior are controlled by what they see, so the first thing I always do is attach cardboard to the open barred sides of the squeeze chute. The cardboard prevents cattle entering the chute from being forced to approach visible people standing close to the chute, where they are deep inside their flight zone.[23, 24]

I also get the people to stop yelling and constantly shocking the cattle with a prod. This keeps the RAGE emotion low, and FEAR probably drops from the level of what a person would feel getting mugged in the subway down to a very unpleasant doctor's appointment. I can prove that FEAR is reduced because cattle poop less in the squeeze chute when they are handled quietly. Behavior researchers use defecation scoring as a measure of fear stress. If there's less poop, there's less FEAR. I tell ranchers that the cattle poop less because they are no longer scaring the s*** out of them.

After I get the cardboard blinds installed, I train handlers to use the cow's *point of balance* at the shoulder. Point of balance movement is another hard-wired behavior pattern. Cattle move forward when a person quickly walks past their shoulder in the opposite direction of the desired movement. When handlers use the cow's antipredator software instead of electric prods and yelling, cattle struggle, run, and poop less.

To completely eliminate FEAR, handlers have to train their animals to voluntarily enter the restraint device to get a tasty treat. Habituation and positive reinforcement are key. Progressive ranchers walk their calves through the handling chutes to habituate them and reduce FEAR, and reinforce them for accepting restraint. If you're planning to show cattle at the county fair, of course, you have to habituate them to bikes, flags, balloons, and other novel things they are likely to see there.

The PANIC System: The Social Emotions of Cattle

For cattle, the major welfare issue involving the PANIC system is abrupt weaning of calves, which is extremely traumatic and should never be done. When calves are weaned abruptly, the moms and babies are separated six months after birth, and the calves are allowed to bawl and scream and pace for three to five days trying to get back to their moms. Sometimes ranchers put the calves on a truck and ship them together in a big bunch with all of them bawling and screaming, which is an even more terrible thing to do.

There are two ways to wean the babies from their mamas without traumatizing them. One is *fence-line weaning*, where you put the mom on one side of a fence and the baby on the other side. The mama and the calf can still stand close together, which is what the calf wants, but the calf can't nurse. If you do this forty-five days before you ship, the calves are fine. The mama and calf stand alongside each other for three or four days and then naturally drift apart. There's a little bawling on both sides of the fence but not much.

The other approach to low-stress weaning, which may be even better,[25] is a two-step procedure where the calf is weaned several days *before* being physically removed from his mother. To cut off the milk supply without separating the calf from his mama, the stockperson inserts a plastic tab called an EasyWean into the calf's nostrils. It works like a clip-on earring and doesn't hurt the calf. The tab hangs down over the calf's mouth and keeps him from getting his mom's teat between his lips. The main disadvantage of this method is that the calves have to be handled twice in the squeeze chute, once to put the tab on and again to take it off. This requires extra labor. On ranches where the calves have to be handled twice for vaccination, this method works well.

Derek Haley at the University of Saskatchewan did his doctoral research on two-step weaning.[26] When he weaned a group of calves four days before separating them from their mothers, he found that the mamas vocalized 80 percent less than cows whose calves were abruptly weaned. The calves vocalized *95 percent less* than calves weaned in the traditional high-stress manner. The two-step-weaned calves vocalized at about the same frequency after weaning as they did before weaning.

Sudden weaning is completely unnecessary, and people need to be encouraged to switch over to low-stress weaning. Abruptly weaned calves have reduced weight gain for a week and higher stress levels.[27] I hope as word gets out that abrupt weaning is as bad for production as it is for the calves that all ranchers will drop the practice. Unfortunately there are some ranchers who still wean their calves the day they transport them.

The welfare issue for dairy cows is different. A dairy cow has to have a calf every year to keep producing milk, and she can't nurse her calf because she has to go back on the line. The calves are taken away on the day they're born and raised individually in stalls, where they live for six to eight weeks before moving to group pens. The weird thing is that you can go to these calf farms and you don't hear all the bawling and screaming you'd hear on a beef farm. A beef calf would go nuts alone in a hutch, but Holstein calves just don't get that upset being taken away from mom. That's because we've bred some of the social bonding out of them. The reactions of the PANIC system are less intense. The Holstein mama is less traumatized than a beef cow when she loses her calf, too. She's been bred so intensively for milking that she doesn't care as much about her baby. A beef cow would be screaming if you took her newborn baby away.

The big welfare issue with dairy cows is: What do you do with all the calves? It would be better for the calves socially to be raised with other calves, and some ranchers are doing that now. But newborn Holstein calves are so fragile from a disease standpoint that it's hard. Some ranchers put the newborn calves in hutches where they can see each other but not touch; other ranchers put them alone in stalls with open sides so they can touch each other but still have enough separation to keep from getting sick.

Cattle Groupings and Regroupings

The other unnatural thing that happens to both dairy and beef cattle is that they get artificially grouped and regrouped a lot for transport, for fattening up at the feed yards, and for different milking groups. Regrouping isn't traumatic for cows but it *is* stressful, probably for the same reasons it would be stressful to people. Cows prefer being with the cows they already know and don't especially like being thrown together with strangers. Anytime cattle are regrouped you see some hostile behaviors, mostly in the form of pushing, butting, and chasing. Newcomers have a harder time than the cattle that were already in the group.

To prevent intense fighting between animals, ranchers and feed yard operators should follow these behavioral principles:

1. *Calves should be raised with other calves, not in isolation.* Isolation-reared cattle are more aggressive and less able to adjust to new social groupings.[28] All animals have to be socialized from a young age, and cows are no exception.
2. *Groups of cattle should be no smaller than four.* In many species larger groups are more peaceful than smaller groups, and cows have been found to be less aggressive toward each other in larger groups, too.[29] One of the reasons bigger groups work better is that they are usually housed in larger pens, and an animal that is attacked can move away. In pigs I have observed that four or five

strange pigs mixed together will fight more than a group of one hundred strange pigs.

3. *Larger is better, but groups can probably be too big, too.* Cattle researchers Joseph Stookey and Jon Watts recommend that no more than two hundred cattle be put together inside one pen.[30]
4. *If possible, young dairy heifers should be regrouped a few times to get them ready to join the adult herd.* One study found that the optimum number of regroupings is seven. In that study heifers were regrouped once or twice weekly, sixteen times. The heifers never got completely used to being regrouped; each time the group changed there were aggressive interactions with the new group mates. But they were least aggressive on the seventh regrouping and most aggressive on the sixteenth.[31] After seven regroupings, the cattle probably got really stressed, which led to more fighting.
5. *If you can keep cattle together with some of their buddies when you regroup them, that's best.* Cows are calmest when they're with cows they know.[32] Although cattle don't especially like being regrouped, they adjust quickly. By the end of two weeks the aggressive behavior is gone and the cows have formed new social attachments.[33]

Bullies and Bullers

The one horrible exception to this rule is the *buller steer syndrome* where a bunch of *rider* steers gang up on another steer — the *buller* — and mount him over and over again until he's exhausted and broken down. The buller has hair loss, swelling and injury on his rump, and even broken bones in some cases.[34] Sometimes a buller is ridden to death. It's disgusting.

Nobody knows why bulling happens, or why there are many pens of cattle where it never occurs. The two factors most strongly associated with buller syndrome are the steroid hormones given to steers to make them grow faster and the degree of crowding in the pen. Unpublished computerized records from many feed yards showed that bullers were more prevalent in pens of four hundred cattle compared to pens of two hundred.

Many researchers think bulling starts when cattle fight over resources such as water and then progresses to a weird sexual behavior for reasons we don't understand. That's possible. I have seen a dominant steer stand alongside a water trough as if he owned it, although I don't know whether that steer turned into a rider. There is also an element of observational learning, so it is important to remove the buller promptly before more and more steers learn to attack it.

Joseph Stookey and Jon Watts recommend that facility designers create pen designs that can protect bullers and that group size be limited to two hundred. Adding more water troughs might help, too. The records I just mentioned showed that the four-hundred-cattle pens had only two water troughs, the same as the two-hundred-head pens where there was less bulling. Jon Watts also thinks that buller syndrome may be a sign that *many* of the cattle in a large group are suffering, not just the buller: "The proportion of bullers in a group might represent merely the most extreme manifestation of a larger welfare problem, chronic social stress in very large groups."[35]

No one has a solution to the buller problem, which, besides being horrible for the victim steers, is very costly to the industry. Until researchers find out what's causing it, the only thing anyone can do is transfer bullers to a hospital pen and treat their injuries and infections. Then the bullers have to be kept in a separate buller pen. They can't go back to the pen where they were ridden.

Right now the two top candidates for an explanation are increased numbers of cattle in a pen with fewer water troughs and excessive use of the hormone implants cattle feeders give steers to make them grow faster. Possibly one type of implant causes more problems. Another factor may be sloppy handling of the implants. The implants are slow-release pellets, which a stock-person inserts under the skin of the steer's ear. If the pellets get crushed during insertion, all the hormone is released at once. The buller problem is difficult to study because only a few cattle out of hundreds have a problem. To get answers will require statistical analysis of data from thousands of cattle.

Cows Like to SEEK

Cows like to learn new things. Two researchers did a wonderful experiment on cattle's "emotional reactions to learning."[36] They used a *yoked design* in which one cow had to learn a task in order to get a reward while the other cow got the exact same reward without doing anything to earn it. Whenever the student cow learned the task and received a reward, the yoked cow received a reward, too. Because both cows got the same reward but only one cow learned something new, the experimenters could look to see whether the cow that was doing the learning was happier or more excited than the cow that wasn't.

The task was simple. The student cows had to press a panel to open a gate that released them into a fifteen-meter-long "race," or chute, that led to a food trough. The experimenters measured the cattle's heart rate and speed of movement down the race to see how excited they were.

They found that the student cows were much more excited on days when their learning had improved. The yoked cows, the ones that didn't learn anything new, never showed any spike in heart rate or speed of movement down the race. Learning how to get a reward turns on SEEKING.

Cows that live on the open range can satisfy their SEEKING emotion. The cows I'm concerned about are the dairy cattle. Many dairy cows live their entire productive life inside a barn and never go outside. On some farms they sleep on waterbeds and the only exercise they get is short walks to the milking parlor. They have nothing to explore and nothing to do with their minds.

Poor physical welfare often goes along with poor emotional welfare. Lameness on a modern dairy ranges from 5 percent at well-managed dairies to over 50 percent at the worst dairies. My graduate student Wendy Fulwider surveyed 113 dairies and found that there were big differences in leg injuries. There was no difference between large and small dairies. Well-managed dairies that bedded the cows more frequently with either sawdust or straw had fewer leg injuries.[37] The humans did not know how poor their animals' welfare was. When the researchers asked producers, "Are cows better off today with regard to welfare than they were twenty years ago?" 70 percent said yes. Their attitude was that life was better because being locked up in a dairy stall was like living in a fancy hotel with room service. Only 8 percent thought that conditions for cows were worse, and that cows should have pasture.

The other welfare problem with dairy cows is that breeders have been pushing their biology too hard. There's a huge difference in strength between baby Holstein and baby beef calves. The little Angus calf will be up and nursing and running with mama a few hours after birth but the Holstein calf is not fully mobile for two days. Breeders have overselected so much for milk production that they've created a weak, fragile animal that's so frail it's starting to be hard to breed them. Holstein cows can carry a pregnancy to term but it's hard to get a pregnancy started.

Also, some producers are feeding Holsteins too much grain instead of roughage to force them to grow faster than they should. They don't give the heifers enough time to grow a solid skeleton and hard hooves that would make them less susceptible to lameness. It takes two years for a heifer to grow up and on some dairies she lasts for only two years of milking, then she has to be slaughtered because she's sick or lame. If you think of the cow's metabolism as being like the rpm tachometer in a car, the industry is running dairy cattle in the red zone and burning them up. Ranchers don't do this with beef cows because beef cows live out in the country where they need strength to walk long distances and protect their calves from predators. Beef cows are not as highly genetically selected as dairy cows.

Seeing the grass-fed dairy Holsteins when I visited New Zealand was interesting. They are stronger, beefier-looking animals and lameness is seldom a problem. Since they live outside all year they are less genetically selected to be nonstop milking machines. Physically and emotionally, pastured dairy cattle are in better shape than dairy cattle that always live inside barns and never go outside. Housing has little effect on the quality of the milk because all dairy cows are fed large amounts of forages or silage.

Why People Keep Doing Things That Don't Work

During thirty years of work on livestock handling and the design of restraining devices for animals, I have seen that many people try to restrain animals using force instead of behavioral principles. Even when plants know they're losing money by shocking and yelling at the animals, they still do it. In one slaughter plant I documented a $500 to $1,000 savings per day after I had trained employees to handle cattle quietly, but when I left, workers quickly went back to their old rough ways. Since rough handling doesn't work very well and is terrible for the animals, why do people keep doing it?

Part of the answer is lack of knowledge. When I visited large farms, I was shocked to discover that the managers of many farms run by corporations were often not aware of most of the animal behavior research.

Even when managers do know something about animals, change is not easy. On websites about low-stress cattle handling you can find ranchers talking about how difficult it was to master the new methods and apply them consistently. When I've given talks at cattle conferences, I have had many ranchers' wives say to me, "I wish I could get my husband to stop yelling at the cattle." Replacing old bad habits is hard.

Another obstacle is that to be a good stockperson you have to recognize that an animal is a conscious being that has feelings, and some people don't want to think of animals that way. This is true of researchers and veterinarians as well as stockpeople. I've found a certain percentage of veterinarians and physiologists who deny feeling and emotions in animals. An animal can be violently struggling and crying, but if its heart rate hasn't gone up these professionals insist that the animal is not distressed. Would they say the same thing if they measured a low heart rate in a person screaming in pain because the dentist's drill just hit a nerve? No. They would see the whole person as more important than some readings on an instrument.

Too many researchers today see an animal as a kind of living machine made up of a lot of little chemical parts. They're looking at tissue under a microscope and they're a zillion miles away from the cow. When I listen to long technical lectures about cow hormones at conferences I want to say, "There's an animal attached to that ovary." Researchers need to look at the whole animal.

People working in management often do not want to find out that a widely used agricultural practice is stressful or painful to an animal. Researchers have told me that funding for research whose results could force a change in agricultural practices can be difficult to get. Management's policy is *See no evil.*

Another big factor is turnover in personnel. The normal quit rate in U.S. business is 15 percent per year, and it may be a lot higher on farms and ranches. Grahame Coleman found that 50 percent of new pig stockpeople on Australian farms quit over a six-month period, and he says that the U.S. rate is probably the same based on anecdotal reports.[38] If half your work force quits every six months, you're going to need a very strong training program for new hires, and you *must* have frequent auditing of performance on the job.

Last, very often people find positive handling methods harder to use than negative methods. The blue-ribbon emotions help us understand why. Handling untamed, untrained cattle is frustrating because they don't do what you want them to do, and frustration is a mild form of RAGE. So, unless a person is an expert in quiet handling of cattle, the environment at a ranch, a dairy farm, or a slaughterhouse will naturally activate the RAGE system in his brain. That's why it's easy for people to blow up at farm animals (or at small children). Getting angry at frustrating situations is natural.

Individual personality and temperament determine how intensely a stockperson's RAGE system gets activated by wild cattle. One study found that dairy cattle managed by stockpeople who were "introverted and confident" had higher milk yields.[39,40]

Confident people have more positive emotions than depressed and insecure people, which might mean that their SEEKING system is activated. Since SEEKING inhibits RAGE, maybe confident stockpeople have a higher frustration tolerance. The reason why the first study found that introverted handlers had the most productive cattle is probably that introverted people are naturally quieter than extroverts. Cattle prefer quiet handling.

Managers should hire people with confident personalities whenever they can, and they should improve the working environment by building in as many rewards as possible for good treatment of the animals. This has been done in the poultry and pork industries in some cases, but it needs to be done for cattle, too. The reason incentive programs are less common in the cattle industry is that there is less vertical interaction when the same company owns the animals from birth to slaughter. The cattle industry should give workers financial incentives for reductions in bruises, injuries, and vocalizations during handling.

This would be easy to do in large feedlots and meat plants. The worst way to pay workers is on a piecework basis, which gives employees an incentive to vaccinate as many cattle as they can and always results in rough handling.

The other employee factor managers need to consider is fatigue. Unpublished data collected on chicken and pig truck-loading crews showed that after about six hours of work, people became fatigued and injuries and death losses increased. Fatigue means reduced frontal lobe function, and reduced frontal lobe function means reduced ability to regulate emotions. I have observed on many farms and feedlots that tired workers blow up at the animals more than well-rested workers. The worst situation is an understaffed, fatigued crew.

People Manage What They Measure

In 1990 I developed a conveyorized handling system for cattle in slaughter plants called the *center-track restrainer* that is much more humane for the cows than the old system. Everyplace I was hired to install it, I also trained the workers in how to handle the cattle gently. One of the biggest frustrations in my career has been that I'd do an installation at a plant and train the workers and get the handling real super-good, then I'd come back a year later and find they'd reverted to using the electric prod and screaming at the cattle. This is true everywhere — ranches, feedlots, slaughterhouses. People don't maintain the improvements they make. Often they do not realize that they have gradually reverted to their old bad ways. I call that *bad becoming normal.*

The only answer to this problem is to audit animal welfare. A stockperson's job has enough frustrations built into it that there will always be emotional pressure on employees to revert to bad handling even after they've learned how to handle the animals nicely. The industry has to audit animal welfare continuously.

Ranchers need to start auditing welfare, too. At a grazing conference in Canada, I gave a talk for ranchers where one of the things I suggested they do was start measuring their cattle handling using a scoring system, similar to what I used for the McDonald's audit of animal welfare inside meatpacking plants that I described in *Animals in Translation.*

For ranches and feedlots, I suggest:

- electric prodding: 5 percent of cattle or less
- cattle that fall: no more than 1 percent
- cattle that moo and bellow during handling: no more than 3 percent
- cattle that run into gates and fences: 1 percent or less
- cattle that move faster than a trot: 25 percent or less (trotting is fine but I don't want cattle running when they come out of the squeeze chute; it's dangerous and it's a sign of FEAR)

Using Technology to Lower RAGE and FEAR in People

I've spent a lot of my career designing handling facilities for farm animals, so I know what a difference good engineering makes. Cattle move easily through my curved chute designs because my designs take advantage of the natural behavior of cattle wanting to go back to where they came from.[41,42] However, there is no technological substitute for *understanding and working with an animal's behavior.* The equipment I design is all behaviorally based; it will work only if you're handling the cattle properly.

This is a very difficult concept to get across. People adopt new handling equipment much more quickly than they adopt the behavioral principles they need to make a piece of equipment work. I can prove this with sales statistics from my website. I get twice as many orders for $55 books on how to build corrals and races as I do for $59 videotapes on the principles of good stockmanship. People think buying the technology is all they need to do.

This has been a constant problem in my plant installations. Equipment builders would build the parts of the design I created to control the cow *physically* but leave out the parts I designed to manage the cow *mentally.* They would actually edit and revise the blueprints as they went along. A lot of them would leave out the false floor I designed for the entrance to the conveyor belt so the cattle wouldn't balk when they saw a "visual cliff" below the belt they were supposed to ride on. Or they would get rid of the roof I designed to go over the cattle's heads to keep them from seeing an escape route until they were settled down and riding quietly on the conveyor belt. At one plant I persuaded the crew that the roof was necessary by laying a two-foot piece of cardboard across the system. That one piece of cardboard instantly made the cattle calm, and after that the crew built the roof the way the plans said to build it.

This kind of thing happened repeatedly because the crews didn't understand the purely behavioral reason for having an extra piece of metal on an installation that would have to be maintained and cleaned. I went to the start-ups of the first seven systems to make sure all of the parts that were needed to behaviorally control cattle were put back on.

Managing the Emotions of People

When I started out in the 1970s I thought I could fix everything with engineering. I wasn't thinking that much about managing the behavior and emotions of the people. It took me thirty-five years to learn that about 20 percent of employees can maintain good stockmanship on their own, but the rest have to have incentives because good stockmanship is so against their nature. Incentives work and they turn on a person's SEEKING system. One plant gave prizes of $100 to $200 each month to the two truck drivers who had the fewest dead pigs. The prizes, along with a big chart in the office showing every driver's "dead score," motivated the drivers to handle their pigs carefully and reduce death losses. My ultimate technological dream is electronic measurement of cattle handling with automatic payroll deductions or bonuses. The computer would make a deduction when an animal crashes into the front of the squeeze chute or runs out really fast. Crews that are able to handle cattle quietly would get a bonus. An automatic system of financial penalties and rewards is a techno-fix that would work. Today this is only a dream.

There are a couple of other technological aids to cattle welfare I'd like to see adopted by large feedlots that could automatically audit animal handling. Squeeze chutes with pressure sensors built in to monitor squeeze pressure and struggling would be a great innovation. The technology already exists for doing this. I'd also like to see radar cameras installed in feedlots to clock the speed of cattle as they exit the squeeze chute. Cattle that are handled calmly are less likely to run when they are released. The faster a cow moves, the more upset it is, and there is research showing that "speeders" and cattle that struggle violently gain less weight, which will help motivate plants to install this equipment if somebody develops it.[43,44]

Preventing rough handling is like controlling speeding on the highways. You need constant measurement and enforcement. I have observed that some people enjoy abusing animals. Those people shouldn't be working with animals at all. They're like drunk drivers with multiple arrests who get their driver's licenses taken away. But most employees who are handling the cattle roughly aren't sadistic by nature. They just don't have the training and practice they need to manage cattle well enough that their own RAGE system doesn't get overactivated—and they don't have a system of ongoing audits to make sure that they *keep* using good handling techniques after they've learned them.

The good news is that conditions in the plants are much better today than they were in the early '90s. The animal welfare audits required by McDonald's, Whole Foods, and other companies have forced plant management to monitor, measure, and improve employee behavior. Plants are maintaining their equipment better and reassigning or firing employees who abuse animals. Some plants have installed video systems on the plant floor, which solves the problem of people behaving properly when they are being watched and reverting to old rough ways when nobody is around.

It is impossible to handle an animal that is too weak to stand or walk with good stockmanship. It is the responsibility of both dairy and beef cattle managers to euthanize on the farm cattle that are too weak to travel to another location, such as a meat plant or auction. Some of the worst animal abuses have occurred when "downer" cows that could not walk were dragged or beaten. When these abuses occur, I place the blame on the manager who supervises the stockpeople. Over the years, I have observed that severe abuses almost never occur in places with good management.

The stockpeople have to manage the cattle, and the plant and ranch managers have to manage the stockpeople. To manage employees, managers have to design good work environments, they have to provide training, and they have to audit performance.

6 Pigs

PIGS ARE HIGHLY CURIOUS ANIMALS that have to have something to do with their minds and their snouts, which they stick into everything they can reach. Their SEEKING emotion is almost hyperactive, which is probably related to the fact that they are omnivorous animals (they eat plants and meat) and their ancestors spent a huge amount of time searching for food in the wild. One study found that pigs living in a seminatural environment spent 52 percent of daylight hours rooting and grazing and another 23 percent walking around investigating the environment. They are driven to explore their world.

Pigs get into anything they can. If you leave the hose you use for cleaning the pen within their reach, they will chew it up and destroy it. One pen of my research pigs at the University of Illinois learned to unscrew the bolts that held the pen divider to the wall. As fast as I screwed the bolts back in, the pigs put their snouts and mouths on them and unscrewed them again. I couldn't keep ahead of them.

Candace Croney and Stan Curtis at Pennsylvania State University made an indestructible video game joystick by attaching a car gearshift to a standard game controller inside a very sturdy box. (It had to be strong so the pigs wouldn't chew it up before they learned to play.) The pigs quickly learned that they could move the cursor on a computer screen with the joystick. At first, the game was very easy. The cursor was in the middle of the computer screen and the pigs got a treat if they moved the cursor far enough in any direction to touch a line that formed a square around the cursor.

Once the pigs could do that, the game became progressively more difficult. Portions of the square gradually disappeared, so the pigs had to move the cursor in a particular direction to hit a line segment. They could do it. They weren't just doing it for the food reward, either. When the treat feeder broke, the pigs kept playing. Pigs have a very strong SEEKING system. Curiosity and the urge to investigate new things are part of the SEEKING system, which is very strong in pigs.

I don't want to get into speculation on comparative intelligence and IQ, but I will point out that almost everyone who spends a lot of time around pigs ends up thinking they're very smart animals. That's why George Orwell made the pigs the leaders of the revolution in *Animal Farm.* It's probably also why Winston Churchill said, "I like pigs. Dogs look up to us. Cats look down on us. Pigs treat us as equals." The famous animal trainer Keller Breland, who was the first person to use B. F. Skinner's work to train performing animals along with his wife, Marian, told *Time* magazine that pigs were the most intelligent animals they worked with, followed by raccoons, dogs, and cats.[1] Pigs are highly social animals, too. They are affectionate, which is why potbellied pigs have become somewhat popular as pets. In *The Hog Book,* William Hedgepeth says that English peasants used pigs for hunting from the eleventh to the fifteenth centuries because they weren't allowed to keep hunting dogs.[2, 3]

The domestic pigs are modern descendants of wild boars. The wild relatives of the domestic pig are formidable animals. The males have long, curled tusks and they will attack a person who threatens them. The domestic pig can also be vicious. Pigs can even form temporary "gangs" to attack another pig in the pen. If a stockperson doesn't rescue the victim pig, it can be seriously injured or killed within an hour. No one knows why. However, I have observed that animals with large brains are often the worst offenders for organized gang behavior. Adolescent male dolphins sometimes form gangs to rape females, and rival groups of chimps fight wars.

Intensification of the Pig Industry and the Problems for Pigs

Raising hogs for market is not an easy job. Fifty years ago, farmers kept their pigs outdoors in dirt lots with simple lean-to shelters or sheds to protect them in bad weather. This setup worked fine if the soil was sandy and the drainage was good, and it was good for the pigs' mental welfare because they were physically free and could root around in the soil and mud. But there's not very much sandy soil in the Carolinas and the Midwest, where most pig farms are now located. Pigs' natural rooting behavior combined with normal rainfall and snowmelt made traditional pig farms into ankle-deep mud-pie messes half the year unless the pigs were housed on large pastures.

The solution farmers came up with was to move their pigs onto concrete lots enclosed in open-sided or three-walled structures. That got rid of the mud problem, but it created a new problem, which was removal of the manure and straw. Shoveling out the lean-tos was such a big job that it limited the number of pigs a farm could handle.

So, to fix that problem, the small farmers in Iowa and Illinois started building fully enclosed pig barns with automatic feeding troughs and slatted floors to let the pigs' manure fall into a gutter or a big storage pit underneath the facility. Farmers who installed the new system saved on labor because they didn't have to hire people to remove the manure, but the new barns were expensive, so farmers had to raise more pigs to make a profit. Another benefit of bringing pigs indoors is the prevention of trichinosis, because the pigs are less likely to eat infected rodents.

That created a whole new set of problems because farmers started to have litters of pigs year-round instead of raising new piglets outside during the summer. A pig pregnancy lasts three months, three weeks, and three days (that's an old pig farmer's saying), so with early weaning of the piglets a mama pig can have at least two litters a year.[4] The winters in the Midwest are brutal, and pigs born in the winter could be lost in snowdrifts in the old system, so the mama pigs had to be kept inside. This led to the invention of *gestation stalls* where a sow is kept confined during her entire pregnancy. The sow can lie down and stand up, but she cannot turn around. It's like being stuffed into the middle seat of a jam-packed jumbo jet for your whole adult life, and you're not ever allowed out in the aisle.

Also, early weaning means you have to build more expensive nurseries and provide expensive feed to take the place of the sows' milk, which produces more pressure for increasing productivity.

Basically, every time the pig industry comes up with a solution to a problem, the solution costs so much to implement that the industry has to intensify production — raise more pigs on the same amount of land — to stay profitable. Then the things they do to increase productivity create new problems that need new expensive solutions. The industry has gone through a fifty-year cycle, which is still going on today, of expensive improvements that require increased productivity to be profitable.[5] Most of these improvements have lowered the emotional welfare of the pigs.

Sow Stalls and the Blue-Ribbon Emotions

Most commercially farmed pigs are bored and lack stimulation, but sows locked up in the sow stalls are in the worst condition.

The stall activates the RAGE system when a sow is first put inside because it is a severe form of restraint, which frustrates the animal. I have seen young sows the day after they were first placed in a stall. Their rear ends were all covered with poo and several had injured their tails trying to back out and escape. All animals need to move and are motivated to move, including pigs. Jim McFarlane and Stanley Curtis at the University of Illinois put two groups of sows in stalls large enough to allow them to turn around. One group of pigs had its feeder and waterer both located at the same end of the stall; the other group had its feeder and waterer placed at opposite sides of the stall. Both groups turned around about the same number of times every day, even though one group didn't have to turn around at all.[6]

Sow stalls also increase FEAR. One study that compared sows living in sow stalls to sows living in large groups found that the sow-stall pigs were more afraid of the experimenter than the group-housed pigs, probably because sows in sow stalls have less positive contact with the stockperson. The sows in the group-housing system interacted with the stockperson when he walked through their pens every day.[7]

Possibly the worst aspect of sow stalls is that they starve the pigs' SEEKING system. A sow locked up inside a sow stall has nothing to do with her mind or her snout. Not enough SEEKING is a bad thing in itself, but in research I did at the University of Illinois I found that a SEEKING deficit also increases FEAR, which is what Dr. Panksepp's work predicts. Dr. Panksepp says that SEEKING inhibits fear, so if you reduce opportunities for SEEKING, you're likely to increase the sensitivity of the FEAR system. In my study, I had groups of six pigs in pens with concrete slats. One group of pigs lived in plain pens; the other group lived in pens where people came in and out and where the pigs had rubber hoses to chew. That group also took weekly walks in the aisle between the pens. At the end of six weeks, we tested fearfulness by having a strange person suddenly jerk open each pen gate and stomp into the pen. The pigs in the plain pens were much more likely than the pigs in the stimulating pens to go berserk and run screeching to the other side of the pen.[8] For those pigs, FEAR definitely suppressed SEEKING.

If SEEKING reduces FEAR, then why were the cattle described in the previous chapter acting curiously afraid when they alternated between SEEKING and FEAR? The SEEKING system motivated the cattle to approach a novel clipboard lying still on the ground. When the wind suddenly moved the paper, they jumped back. When the novel paper suddenly moved, it turned on FEAR, but when the paper stopped moving, the SEEKING system took back over. The SEEKING system has the power to reduce fear, but it does not totally shut down the FEAR system. The FEAR and SEEKING systems may operate like different-sized weights put on the opposite ends of a balance scale.

Sows in stalls don't have their social needs met, either. Pigs are social animals that don't like being alone. In the wild, pigs live in small groups and probably survive by hiding from predators. Pigs need to interact with other pigs, and just lying beside another sow in the next stall probably does not satisfy social needs. So the stalls may be bad for the PANIC system, too.

The stalls may also have contributed to the decline of the pigs' physical well-being. I know this isn't a book about physical welfare, but I want to say something about the physical problems of sows in sow stalls. The sows go lame at very high rates inside the stalls and their bones weaken. They go lame partly because they don't get any exercise, and partly because they move so little that stockpeople can't see the first signs of developing lameness. There's no way to tell that an animal is going lame if you can't see the animal walk.

Sows have so many genetic problems now that they're living only long enough to bear three litters. A sow without genetic problems could have twice as many litters. When an animal is kept in a box, breeders forget about breeding for important traits such as strong feet and legs. I am appalled at the foot and ankle defects I have seen. In some young pigs, the ankles had collapsed and the pigs were walking on their dewclaws, which are the two little nubs on the back of the leg above the hoof. This happened because breeders were single-mindedly selecting for fast growth and high weight. Fortunately, some breeders are adopting a more whole-pig approach to breeding and are finally changing the genetics to correct these serious defects.

The industry needs to get rid of sow stalls.

Sow-Stall Alternatives That Work

Unfortunately, the industry continues to prefer hard technological solutions to soft behavioral or management solutions. Keeping sows locked up alone saves on labor and training because it takes fewer employees and a lower level of skill to manage sows in sow stalls than it does to care for sows living in pens. For years I have said it takes a good stockperson to manage sows in pens, but any idiot can manage sow stalls.

In Europe, there's a popular loose housing system that I'd like to see adopted by other countries including the United States. *Loose housing* means the sows live together in groups, not alone in separate stalls. The European system uses one set of feed stalls for four pens, with thirty or forty sows in each pen. A stockperson opens the gates one pen at a time, and the sows go to a row of individual feeding stalls, then come back to the pen. They're alone only when they eat to prevent fighting.

Large groups work great because there are enough pigs in the group to keep them from constantly fighting for dominance. The worst thing you can do to a pig is to repeatedly mix and remix small groups of strange animals together. Pigs mixed in groups of four just tear each other up, partly because small groups are put in small pens and a pig that is being attacked has no way to escape as it would in the wild or in a larger enclosure.

I've been saying this since the 1980s and it's taken until now for the industry and the scientists to figure it out. I'd tell a pig researcher, "I go into the meat plant and they unload a big group of 150 hogs with strange pigs all mixed together. There's hardly a fight," and the researchers would say, "That's just anecdotal evidence." I'd tell them that there's a point where anecdotal evidence becomes truth. Mixing just three or four animals in a small pen is like putting scorpions in a bottle.

I think one of the reasons sow stalls were adopted in the first place may have been that when farmers brought their pigs indoors, some of them made the mistake of putting their pigs together in groups of three and four and repeatedly remixing them. If they'd had their pigs in large groups from the start, they wouldn't have had all the problems with aggression that they did, and maybe they wouldn't have created the need to lock the pigs up alone in cages to prevent it. Work with the animal's natural behavior: That message has to be repeated over and over again. Small groups work only when the sows are seldom remixed. Today, some farmers are successfully using small "static" groups, where the sows are kept in the same group through multiple pregnancies. I hope other farmers will follow their example.

One thing industry workers are right about: We can't improve welfare just by letting all the sows out of their stalls and putting them in pens. If we did, they would tear each other apart. The problem is genetic. On an old-fashioned dirt farm, aggressive pigs would be culled before they reproduced, and the genetics of aggressiveness would be kept in check. But since sow stalls artificially prevent aggression, farmers don't know which pigs are aggressive and which aren't, and there hasn't been any culling pressure against aggression for twenty years. Why would mixing sows cause more fighting problems than mixing market pigs? There are two reasons. First, sows are kept on a diet where they are not allowed to eat all they want to prevent obesity. This makes them hungry all the time, which motivates them to fight over food. Second, some genetic lines of lean sows are especially vicious when they fight over food.

I received an e-mail from a producer who said he had to change the pig genetics he was using after he tore out his sow stalls. Some genetic lines of lean, rapidly growing pigs are mean, nasty fighters. He had to replace these pigs with calmer sows that had stronger feet and legs so they could walk around their pens.

To eliminate sow stalls, the industry will need to cull the most aggressive sows. The only way to do this would be to start the conversion from stalls to group housing and keep a few sow stalls for jailing "criminal" aggressive pigs. After the criminal has her pigs, she would be culled and none of her offspring would be kept for breeding.

The PANIC System and Weaning

One area where the emotional welfare of pigs has improved is weaning practices. Up until ten years ago, the industry took babies away from their mamas at ten days of age. That was really bad. It was too early to breed the sows back and the piglets were very upset. Also, the babies would get into fights because early-weaned piglets nose each other's bellies the way they would nose their mama's belly when they want to nurse.[9] Piglets don't like being nosed by other piglets, and they fight back.[10]

Early weaning is bad for piglets physically, too. Since they're not ready to go on solid food, they don't eat enough for the first few days. That may have been some of the motivation for farms to wean later. Farm profits depend on fast and efficient growing, which is why people keep breeding faster and faster growers. Anytime an animal eats less than it should there is a financial loss.

U.S. farms are weaning pigs at three weeks now. One study of gradual weaning in sows and piglets kept outdoors found that by the sixteenth day after birth the sow was "well into the process of weaning the piglets." The authors based this on the fact that at sixteen days after birth, which is when the study began, the sow was terminating all feedings, even though the piglets hadn't yet started eating the solid food put out in troughs for them. They naturally started eating solid food five days later, at three weeks of age.[11] So the U.S. industry weaning standard, weaning at three weeks of age, is probably fine.

The one way in which the weaning process could still be improved would be to follow the advice of a 2007 review of research on weaning distress and not put piglets through too many changes all at once.[12] On a lot of farms, weaning means that the piglets are separated from their moms, put on solid food, taken to a new physical location, and put in with strangers, all on the same day. It's better to do one thing at a time. Let the piglets start on solid food before you take them away from their mamas and put them in a new place with new pen mates. For all species it is best to leave the young in a familiar pen when the mother is removed.

The PANIC System and Grouping

Pigs being raised for slaughter have fairly good emotional welfare when it comes to the PANIC system. They always live in groups with other pigs, which is the way it should be.

The one big challenge of keeping sows for breeding is grouping practices. *All* pigs fight at first when they're mixed in with stranger pigs. The females fight just as much as the males.[13] One to two days after being mixed together, almost half of the sows in a pen will have scratches on their shoulders. Usually the scratches are superficial and heal on their own in a couple of weeks. By four days after mixing the group will have settled down and you don't see any more fighting.

Even though the injuries aren't terrible, there are things the industry can do to decrease the amount of fighting. Dr. Tim Blackwell, who is at the Ontario Ministry of Agriculture and Food, has a list of methods to decrease fighting in groups of newly mixed sows in particular:

- Group the sows according to size.
- Include one boar in the group.
- Give them an extra feed as soon as you've mixed them.
- Mix them late in the day and turn off the lights.
- Put lots of straw, hay, and toys in the pen as a distraction.[14]

A few studies have found that mixing piglets before they're weaned reduces fighting when they're removed from their mamas. However, one study showed that the piglets just transferred their fighting to the preweaning period, which might be worse than having them fight after weaning when their immune systems are stronger.[15]

One interesting thing about that study: The researchers mixed the pigs by taking off the backs of three side-by-side farrowing stalls. The farrowing stall is the place where the pig is taken to give birth. It's bigger than the sow gestation stall, and it contains a small, roofed enclosure called a *creep box* where the babies can go to lie down. The creep box protects piglets from being crushed by the sow and is warmer than the farrowing stall because the piglets lie so close together inside that their body heat raises the temperature.

In the study, when the researchers removed only the backs of three side-by-side farrowing stalls, that created an open corridor outside the stalls, which were still totally separate otherwise. The open backs let the pigs mingle in the corridor outside the farrowing stalls and go inside each other's farrowing stalls. The only place they got into fights was in the corridor, away from the sows. When piglets went inside one of the other farrowing boxes, almost none of them fought even though the piglets were nursing each other's mamas and going inside each other's creep boxes. You'd think the piglets might want to defend their territory, but they didn't.

This reminds me of what I wrote in *Animals in Translation* about the "boar police." The boar police are male pigs that keep the young male pigs in line just by looking at them. The younger pigs won't fight as long as the adult males are watching them. In this study, the mama sows probably also kept the baby pigs from fighting.[16]

A progressive farmer I know mixed nursing piglets by taking down the side partitions between the sows, and the system worked much better. When the piglets left their mama's pen, the only place they could go was directly into another pen with another sow. It's probably not a good idea to have baby piglets mixing alone together out in a corridor. Grouping piglets by sex also reduces aggression. When male pigs from different litters are mixed together with females, they get very aggressive.

The Curious Pig: SEEKING

Pigs have lively, active minds, and they need to live in an enriched environment that lets them stimulate their SEEKING emotion. In my research, the piglets raised in a barren plastic pen were much more hyper than piglets raised on straw. The piglets in the barren pens were also greater stimulus seekers. When I cleaned the pens with a hose, they bit madly at the hose and at the water stream. When I cleaned the feeders, they excitedly bit at my hands. They were starved for stimulation.

Understimulated pigs will chew off each other's tails, too. It's horrible. It might be just one pig that starts it, but once one pig draws blood other pigs can get involved, too. It's not really aggression; the pigs are just desperate for something to explore and chew.[17] Some pigs bred for lean meat have an especially high SEEKING drive and they will root and chew up a person's boots. They also do a lot more tail biting.

Pigs have such an intense SEEKING emotion that good stockpeople use pigs' natural curiosity to move them. Usually, stockpeople think the way to herd pigs out of a pen is to get behind them and drive them forward by pressuring their flight zone. But very often this doesn't work; they just scatter and pile up because people try to push them too quickly. The best way for a stockperson to move pigs out of a pen is to open the gate and just stand right next to it inside the pen. The most curious pigs will come up to investigate the stockperson. Then, after they're done investigating the stockperson, they'll walk out of the pen into the alleyway to investigate *it*. The second most curious pigs will do the same thing. They'll walk up to the stockperson to explore him and then walk into the alleyway to explore the alleyway. At some point the herd instinct will kick in and the rest of the herd will follow the leader pigs out of the pen.[18]

There's another problem with moving pigs, which is that when the pigs are driven they often refuse to walk on an unfamiliar floor surface. I consulted on a farm where piglets that had been raised in pens with plastic floors were impossible to drive down a corridor with a concrete floor. It was utter frustration for the handlers. I suggested opening all the gates and then going out for dinner. When we came back, the pigs were playfully exploring the concrete floor and were now easy to move into the next building. When the piglets were pressured to walk on a concrete floor for the first time, FEAR took over and blocked the urge to SEEK. With the pressure gone, SEEKING overrode FEAR, and the pigs explored the new floor and got used to it.

SEEKING Straw

Pigs are obsessed with straw. When I threw a few flakes of wheat straw into my pen of piglets, they rooted in it at a furious pace. After the straw had been chewed up into tiny short two-inch pieces, they lost interest. The chewed-up straw was now boring and no longer novel.

So far, no one has found anything that can compete with straw for a pig's interest and attention. One recent study tested seventy-four different objects — things like a ball with a bell inside, hanging buckets, pieces of carpet, cloth strips, a metal colander, compost, shredded paper, and sawdust. Lavender straw came in first.

Straw is extremely good for pigs, especially full-length straw. Chopped straw isn't nearly as satisfying.[19] My straw-bedded pigs were calm. When I cleaned the feeders in their pen, they moved quietly away and there was no hyper biting. One study of tail biting found that pigs needed only a small amount of fresh straw two times a day to keep them from biting each other's tails.[20] Straw is so important to pigs that in 2001 the European Union passed a Commission Directive stating that "pigs must have permanent access to a sufficient quantity of material to enable proper investigation and manipulation activities, such as straw, hay, wood, sawdust, mushroom compost, peat or a mixture of such, which does not compromise the health of the animals."[21]

Unfortunately there are two drawbacks to straw. First, we have a limited amount of straw in this country, and we'll have even less suitable bedding materials if we start using straw and chopped cornstalks to make ethanol. Canada has plenty of straw because it grows so much wheat, but it's too expensive to ship Canadian straw to American pig farms.

The solution for limited supplies of straw is to use straw exclusively for enrichment, not for bedding. You need a huge amount of straw to make proper straw bedding so that the pigs will stay clean and not be wallowing in manure. I have seen straw-bedded systems where the farmer skimped on straw and it was disgusting, but you don't need a huge amount of straw to create an enriched environment. All of my pigs were reared on concrete slatted floor pens, and just a small amount of straw made a huge difference in satisfying SEEKING.

The other problem is that in large commercial systems straw clogs up the liquid manure systems because it jams the pumps. This is the reason why most large U.S. farms don't give their pigs straw for enrichment, which they should all be doing. One way to minimize the cost is to "toilet-train" the pigs. When pigpens are set up right, pigs can have very neat pooping habits. They will poop in cold, wet places near a waterer and avoid messing in dry places where they sleep and eat.

If a farm isn't willing to give its pigs straw, it should provide other forms of enrichment. Pigs like odorous objects they can chew, "deform" (meaning the object changes shape or size as the pig manipulates it), and destroy.[22] In the 1980s, I did one of the first experiments showing that pigs like soft things they can chew up and destroy. I hung three objects on strings above a pen of piglets that didn't have straw to root in and chew: a rubber hose, cloth strips, and chains. Each object was attached to a switch that activated a counter whenever a pig pulled on it. The rubber hose and the cloth strips received a lot more votes on the counters than the chains.

Stockpeople also need to rotate the objects they give the pigs. One study says you probably need to rotate toys in and out as often as every two days.[23] My pigs were much more interested in rooting and chewing the new objects I brought each day compared to the objects I had brought just the day before. The instant I dropped an old telephone book into the pen they ripped it to shreds. Then, after they'd fully rooted through all the ripped-up pages, they lost interest in it.[24] Pigs need new things to explore because novelty turns on SEEKING. A pig can't keep exploring the same thing over and over again.

The other approach to satisfying the SEEKING emotion that I like is announced rewards. Instead of just giving the pigs their straw a couple of times a day, you use a conditioned signal to alert them to the fact that the straw is coming. That puts the pigs into the looking-forward-to state all animals and people love. There have been two studies of announced rewards or enrichments that I know of. Both found that an announced reward or enrichment made the animals act happier. One study looked at two groups of piglets that were stressed out because they were being weaned. In the experimental group, the pigs heard a doorbell just before a door opened up and let them into a hallway covered with straw and mixed seeds. That turned the doorbell into a conditioned stimulus, like the bell for Pavlov's dogs or the clicker in clicker training. The control-group pigs were given the straw and seeds with no announcement. The "anticipation" pigs played more and fought less after they were weaned than the control pigs.[25]

Why is it so fascinating for pigs to chew up straw until it is in tiny shreds? I think the pigs' motivation is similar to mine in my childhood days when I spent hours dribbling sand through my hands. As each tiny sand grain flowed through my fingers, I scrutinized how the light reflected off of it. Each particle had a different shape, and the sparkling reflections changed as I varied the flow through my fingers. I think this may be true for pigs because they are so intensely focused on the things they manipulate with their noses. Each little flake of straw is different and fascinating, and the pigs are driven to explore and chew their straw until it's all gone. Both pigs and children with autism are obsessed with the things they like to manipulate.

People and Pigs

I'm very frustrated by the way I see so many stockpeople treating animals. Early in my career, I got a letter from John McFarlane at the Massachusetts Society for the Prevention of Cruelty to Animals. He worked with stockyards on animal handling. He was an old man then — he was retired — and he told me you can go out and train people on farms and ranches and stockyards and get the handling really good, but then the old rough ways return after you leave. I was only twenty-five when I got John's letter, and I didn't believe him.

Now I'm sixty years old, and I know he's right. All through my career I've seen that 20 percent of the workers are good stockpeople. I remember, back in Iowa in the 1970s, a great big fat trucker in bib coveralls with big jowls. He looked like one of the pigs — I called him Iowa Cornfed. He'd get out of the truck and grunt at the pigs and they'd come off the truck just as nice and calm as they could be. It was like he was a pig reincarnated as a person. There have always been stockpeople like him. They remain good stockpeople even when they work in bad conditions.

The other 80 percent have to be trained and managed. I've seen horrible handling. I remember in the 1970s going out to an auction to take photos for a talk I was giving. I saw a trucker kick every single pig going into the truck.

Management is responsible for how the workers treat the animals. Employees who really care about pigs get discouraged if the boss tolerates abusive handling of animals. I have seen many excellent stockpeople get burned out when they worked for a boss who did not appreciate the good care they gave the animals. I recently received a letter about this from a herdsman in one of the barns on a big pig farm:

> My problem is that my peers and employers are not interested in my techniques even though they cannot explain the excellent production in my barn. I am looking for advice on how to deal with my employers or where I can take my expertise. I'm starting to become stressed at work watching how my peers handle animals. I have even tried to show my peers your technique of getting down and seeing what the animal sees, all to no avail.

On farms and in slaughter plants, change has to come from management. I was out on a large bison ranch recently and was very pleased with the handling. Why do they have good handling? Because the boss wants it that way. It's coming down from the top. I have seen this pattern many times. Even back in the bad old days before major customers started doing animal welfare audits of the meat plants, there were always some places that did things right. Every one of these plants or farms had a strong manager who served as the "conscience" for the employees. Many times I heard employees stop another employee from hurting an animal by saying, "You can't do that. The boss does not allow it."

Since 1999 animal welfare audits by major meat-buying customers and greater enforcement by the U.S. Department of Agriculture (USDA) have greatly improved handling and stunning practices in meat plants. The auditors and the inspectors have forced plant management to do a better job of training and supervising employees. Another benefit is greatly improved maintenance of plant equipment. When the programs started, three plant managers out of seventy-five plants had to be fired. Unfortunately there are still a few plant managers who have learned how to "act good" during an audit and then revert to rough practices when the auditor leaves. The only way to solve this problem is to do audits with remotely viewed video cameras. Today the audit by McDonald's is mainly at the meat plants, but pig farms need to be audited, too. No one is auditing pig farms.

Ethology of Humans

How do the people at the top get the rest of their employees to handle the animals correctly? Part of the answer is to understand that people are animals, too. There is such a thing as "human nature," and managers should think about stockpeople and about themselves the way animal ethologists think about animals: as conscious beings who predictably follow the rules of behavior for their species. Instead of relying purely on short-term training programs and employee willpower, managers should start thinking like ethologists and expert trainers.

The most important thing an effective manager needs to do to stay on top of his own behavior is to guard against desensitization to the animals' fear and pain. Over the years I have observed that the managers who do the best job of overseeing animal handling are out in their barns every day, but they are not slaving away moving pigs day in and day out to the point where they get so tired that they don't care. A manager in a distant corporate office is too far removed to care about the animals, but a person working in the trenches can get desensitized, or habituated, to suffering.[26]

The first thing an effective manager must do to take care of the animals is get rid of employees who are bullies. I've seen many times that there is always one ringleader for really nasty cruelty to animals. It's the same principle as playground bullying, where there is often one leader and the rest of the kids go along. Take away the leader, and the bullying stops. On a farm or in a meat plant, the ringleader must be either fired or reassigned to a nonanimal job. He should not be working with animals.

Getting rid of the ringleader produces a positive change in the work environment, which is very important. Learning theory shows that positive and negative reinforcements from the environment create and maintain behavior. A good manager creates an environment that reinforces good behavior by employees. The basic principle is: Make the environment work for you, not against you. Never leave up to willpower and self-discipline what you can do with environment. (This is why diet books tell you not to keep potato chips around. A pantry stocked with junk food is a bad environment for a person trying to lose weight.)

For farm and meat plant workers, a good environment means the farm must not be understaffed with bone-tired, overworked people. Guarding against fatigue is important because stock handling will also have frustrations, and a stockperson needs good frontal lobe function to inhibit his impulses to yell at or shock the animals. Fatigue lowers frontal lobe function.

Stockpeople will also have a better attitude if their employer shows that he cares about them. I have been on pig farms where the owners never repaired the equipment or replaced the old shredded, ripped-up coveralls. Those places often have poor animal handling. On well-managed large pig farms, where workers treat the animals well, the company has repairmen come in to maintain the ventilation system and repair equipment. The workers aren't being constantly frustrated by mechanical breakdowns. In terms of the core emotion systems in the brain, well-managed farms reduce the core emotion of RAGE, which reduces the likelihood of workers lashing out at the animals.

Take the Weapon Away

Another very important step management must take is to get the electric prod out of workers' hands. Handing stockpeople an electric prod to carry around goes against everything scientists know about positive and negative reinforcement. Electric prods create a negative reinforcement loop. Every time a stockperson shocks an animal that's not moving, something bad (a balking animal) goes away (the animal starts moving). The more a worker uses the prod, the more he will be reinforced for using the prod, and so his use of the prod escalates.

Before the McDonald's slaughter plant audits went into effect, some plants were at 500 percent prod use, with every single pig being shocked multiple times. The sound of squealing pigs in those places was deafening, and it was impossible to carry on a conversation in the stunning area. I visited one dreadful plant during the night shift. The workers had about a hundred pigs piled up squeeching and flipping over. Electric prods were being thrown into the squealing pile-up like harpoons, retrieved with an attached wire, and thrown again.

You have to handle pigs gently because they're more excitable than cows. When pigs get worked up they'll jam into a chute or entrance. Cattle will back off and let another animal go by, but pigs just keep pushing forward until they get into a complete panic and they're stuck together in a huge clump. Stockpeople call it "squealing Super Glue." The lactic acid levels in their muscles skyrocket from all the exertion, and that wrecks the meat quality. So you really want to limit electric prod use, and yet stockpeople used to use it constantly.

In many places I have seen people become kinder and gentler when their "weapon" was removed from their hands and was replaced with a flag or a plastic paddle. Instead of *zap zap,* the people pat the animal's rump and say, "Come on, boy." There's no negative reinforcement loop working on the humans, and there might even be a positive reinforcement loop.

One key to making this work is that the prods have to be easy to get to when they are needed to move a stubborn animal. If employees have to walk across the room to pick up the prod they'll start carrying it in their hands again, and you'll be back where you started. On one ranch they built a little rack for the prods on the catwalk that runs alongside the cattle chute where the employees stand. The workers don't carry the prods in their hands, but they can pick up a prod from the rack if they need it. The rule is that the prod is to be picked up only when a flag or paddle fails to move the animal and then returned to the rack immediately after use. That system is working well. The workers don't use the prod nearly as often as they would if they were holding it in their hands.

The industry isn't to the point yet where it can ban prods altogether, because if you do you'll get employees taking a two-by-four with a nail in it or steel gate latch rods to beat the animals onto a truck. There are some parts of facilities where the prods should be banned today. Prods should never be used in the sow-breeding barn. Even if management doesn't care about the sows, they should care about production, and sows that are afraid of people have fewer piglets. Electric prods can be eliminated for moving groups of pigs in and out of group pens, too.

There are some innovative group-handling systems that have made it possible to throw electric prods away at slaughter plants. At one of the big pork plants, the plant manager put an electric prod on a wooden plaque labeled *The Last Electric Prod.* It is now hanging in a prominent place on his office wall. I hope all pork plants will adopt those systems eventually. Today many huge pork plants have these systems, but they are too expensive for small slaughter plants.

One simple method to reduce or eliminate electric prod use is to teach pigs how to drive by walking through their pens on the farm to train them to quietly get up and move when a person walks through their pen. Another good method is to periodically move pigs out of their pen and into the alley so they learn how to move in and out through the gate. A third method is to raise the young pigs that will be used for meat in an Auto-Sort system. In this system, the pigs walk single file every day through a scale to eat in a central "food court." When they reach market weight, the electronic scale directs them to a truck-loading pen. Walking through the scale teaches the pigs to move in single file. When they get to the meat plant they run up the single-file chute to the stunner because it looks like the entrance to the "food court."

Being Too Kind Can Be Cruel

Not all problems with animal welfare come from workers being callous or cruel. With pigs, especially, you can get welfare problems from workers being too tender-hearted! Farm workers can have a difficult time making the decision to euthanize a sick pig that is suffering. I've noticed that the really good, caring stockpeople can't stand to put a pig down, especially the wonderful farrowing managers. I met a farm wife who handles all the farrowing on the pig farm that she and her husband own. She loves working with the piglets and she can't put the little runts down no matter how sick and hopeless they are. The way they handle it is that her husband takes care of it when his wife has left. Tender-hearted stockpeople who can't let go of very sick animals is a fairly important welfare problem that also comes from normal human behavior and emotions.

The reason this is a problem isn't just that the animals suffer; it's that the people suffer, too — and when employees *repeatedly* go through the pain of holding on to an animal and watching it suffer and then finally euthanizing it or watching it die, eventually they're going to become desensitized to animal suffering. That's how habituation works.

To deal with this issue, Dr. Blackwell of the Ontario Ministry of Agriculture and Food recommends that employers create standard protocols for the treatment and euthanizing of sick pigs. The protocol tells every employee what to do when a pig is sick, how long to use the first treatment, when to switch to a second treatment if the first treatment isn't working, and when to euthanize an animal if nothing is working. If an employer has a clear protocol in place for when a sick animal is to be euthanized, stockpeople don't spend as much time on the job watching animals suffer. Also, the stockperson doesn't feel guilty (or not *as* guilty) when an animal has to be put down, because it's not his decision. The "rule" makes the decision, not the stockperson, so the danger of desensitization is lowered that way, too.

The rules for putting down sick animals must not be too strict. I have seen good stockpeople save sick animals that were not hopeless. They need to be allowed to do this. I remember one farm where the really small piglets were placed on towels the employees had brought from home. These piglets recovered and became big market pigs. An excellent cage-free chicken farm had a hospital pen for injured laying hens. The farmer told me that most of the hens recovered and returned to the egg-laying flock. He had such a caring attitude toward animals. If he had been told to just wring the neck of every injured bird, he would have become desensitized. To prevent stockpeople from becoming desensitized, the rules must be strict enough the hopeless and suffering animals are euthanized and the ones that can make a full recovery are cared for. Also, no employee should be *forced* to euthanize an animal he has cared for. The job of putting down animals must always be done by a worker who has volunteered to handle it.

Managers should hire some women, too. I've gone to a number of feed yards now where they've got all-girl crews taking care of sick cattle. They are gentler than all-guy groups and they keep the facility cleaner. Also, the people who excel at taking care of baby pigs are women, and the really good farrowing managers are often women. At the University of Illinois farms when I was a student there were two farm managers, Steve and Diane. Steve was the manager of the farrowing barn and Diane took care of the *breeders* (the sow mamas) and the *finishers* (the young pigs growing to market size). She was the newer employee, and after she'd been on the job for a while she and Steve traded jobs and she became the farrowing house manager. Steve was an excellent stockperson, but the sows really settled down with Diane and they produced well. Diane was just superb with farrowing pigs. She could relate. She'd say, "I've had children. I know how those sows feel. I know it hurts."

Cognitive Behavioral Training for Stockpeople

The Australian researcher Paul Hemsworth says that you can't just change stockpeople's behavior; you have to change their attitudes, too, because attitudes underlie behavior. Dr. Hemsworth has created an interactive computer training program called ProHand that teaches stockpeople *why* low-stress handling is important.[27]

Dr. Hemsworth says that to change behavior you need to do three things:

- Change the beliefs that underlie the behavior.
- Change the behavior itself.
- Maintain the changed attitudes and behavior.

Dr. Hemsworth calls this approach *cognitive-behavioral training*. ProHand includes information about the nature of animals and their stress response. Dr. Hemsworth says another goal of the program is "preparing the person to handle stressful situations and reactions from others towards the individual following change."[28] To maintain positive changes in stockpeople's behavior the program uses posters, newsletters, and follow-up training. This is completely different from a standard training program in any industry.

So far, Dr. Hemsworth has done two studies on the use of the ProHand training program on large commercial pig farms. In both studies stockpeople who used the program changed their attitudes and behaviors toward pigs. The people were now stroking pigs and talking softly instead of yelling and hitting them. The pigs seem to have become less fearful, too, judging by the fact that they acted less afraid of the experimenter than the pigs in the control group.[29] Six months after the second study was over Dr. Hemsworth found that 61 percent of the stockpeople who were trained by ProHand were still on the job compared to 47 percent of workers who hadn't had the training.[30]

How to Keep Handling Good

New behaviors don't maintain themselves. So, once you have good handling on a farm or in a plant, you have to maintain it. To maintain good handling of animals, you have to audit using a numerical scoring system. The auditor counts the number of pigs that are stunned correctly and the percentage of pigs that are moved quietly without squealing, falling down, or being zapped with the prod.

Numerical audits make it easy for management to see whether their handling practices are getting better or becoming worse. They're good for the workers, too. When McDonald's started auditing the big plants, the managers often called each other up to brag about their good scores. I think the audits turn on the SEEKING system. The workers were constantly trying to beat their own score or someone else's score.

The animal welfare audits being used by large meat buyers to audit slaughter plants really work. Today in a large, well-run, audited pork plant you can carry on a normal conversation next to the pig stunner and hear only a few intermittent squeals. In the best plant, the electric prod is used on 5 percent or less of the animals.

In 1999 I was hired to train the McDonald's food safety auditors to do the animal welfare audits. That first year we audited twenty-six U.S. slaughter plants. At first the food safety auditors were skeptical about measuring things such as electric prod use and squealing. When they saw how well the audits worked to improve animal treatment, they quickly became enthusiastic. Two of the auditors became so motivated that they spread the audits throughout the McDonald's international system. During this same year, I also worked with Wendy's and Burger King. Today I teach auditor training seminars and I also work with meat plant management to improve animal handling and design better facilities.

All plants can meet these guidelines. One of the worst things I ever saw was a plant we audited for McDonald's where the workers had eight-inch electric chains hanging down onto the pigs' backs from the bars over the chute. When I watched how they were using the chains, I realized they had created a remote-control electric prod system for moving the pigs. They could turn on the current from a row of buttons on a control panel as the pigs went through the chute, just like they were playing the piano.

After I saw that I went into the conference room with the other auditors and we told the plant management, "You just flunked your audit." Then I said all that electric crap had to come out and go in the garbage dumpster and if any of it was there when we came back they would be delisted and could no longer supply McDonald's.

They were sitting there complaining that there was no other way to keep the line flowing. I said, "Oh yes, there is. Your sister plant's doing just fine without electric chains." Their sister plant was a plant in another city owned by the same company. "If your sister plant can do it, you can do it, and you can do it with the chute you already have."

They knuckled down and they did it. When we came back everything was working fine and the chains and the control panel were gone. They did not have to make a single capital improvement to get the pigs moving quietly; they just had to eliminate a lot of little distractions that were scaring the pigs and learn to move smaller groups of pigs more quietly. But without the threat of losing McDonald's as a customer, they never would have done it. Even if they had, they wouldn't have maintained the improvements.

To maintain high standards, the audits and oversight by major customers must never be relaxed. If the audits stopped, 20 to 30 percent of the plants would quickly lapse back into old rough ways. About a quarter of them would stay good because they've always been good. But the rest would regress.

Pig Welfare and Genetics

I went to a plant recently where I saw pigs that were too weak to walk two hundred feet out of the stockyards and up to the stunning area. They just lay down on the ground and didn't move. The stockpeople had to take them off to a separate area and shoot them with a captive bolt stunner instead of stunning them in the stunning area. They had five people in there full-time just to handle these very weak pigs.

The problem is a combination of three things:

- Genetic selection for rapid growth — breeding only the fastest-growing pigs to get offspring that grow even faster. Many of these pigs are lame and have poor leg conformation.
- Growing pigs to heavier weight — in the 1970s, market pigs weighed 220 pounds; today they weigh 275 pounds.
- Feed additive — to give customers leaner meat with less fat, producers feed pigs an additive called Paylean. Paylean makes pigs leaner but it also makes them hyperactive and weak when given in too high a dose.

Breeders need to realize that they can't have everything. There's a saying in engineering: You can build things cheap, fast, or right, but not all three. You have to pick two. It's true with genetics too. You can't breed a pig that grows fast, gets heavy, and stays strong. If you want a heavy pig you have to give it more time to grow; otherwise it will be weak or lame. You have to pick two, and one of them should be "stays strong." Breeding fast-growing, super-heavyweight pigs that are too weak to walk two hundred feet is wrong.

Animal Welfare, Technology, and Building Contractors

The tragedy of sow stalls is that they might never have become an industry standard if the first electronic sow feeders developed in the 1980s hadn't failed. When pig farming moved indoors, some farmers installed computerized electronic feeders that automated the feeding process. Individual pigs wore transponders around their necks that told the computer who they were. If the pig hadn't eaten its full allotment of food for the day, the feeder gate opened and the pig walked inside a small, enclosed feeding area. Then the system delivered an exact amount of food to the trough.

Computerized feeders saved a lot in labor costs, but the new feeders created bad problems with aggression because a pig would walk inside the feeding area to eat, and then have to back out into the faces of the waiting pigs when it was finished. The waiting pigs would bite the pig on the butt or in the privates and some of the injuries were really bad.

The other problem was that electronics hadn't been miniaturized yet, so the sows had to wear a great big huge transponder the size of a tennis ball on a chain around their necks to signal the gate. At one point the company was putting the transponder on seat belt strapping and the sows were walking around with big seat belts around their necks. Cows can wear a necklace and they're fine with it, but pigs chew them up. Some of the pigs figured out that the transponder was the key to the gate, so they'd pick up a chewed collar from the ground and carry it over to the gate and get to eat double rations while the other pig went hungry.

The secret to making the electronic feeder work was to build an exit gate at the front of the feeder so the sow didn't have to go out the way she came in. Computerized feeder systems with two gates work beautifully, but by the time engineers came up with a two-gated design it was too late. The industry had rejected the automatic feeder and they built sow stalls.

A huge amount of the blame for the spread of sow stalls throughout the pork industry goes to the building contractors who built the sow stalls in the 1990s when there was a gigantic expansion phase. Sow stalls became the new industry standard when two start-up hog companies hired contractors who specialized in swine facilities to build hundreds of new barns. The contractors advised the new companies to build sow stalls. The stalls are horrible for the sow but good business for the contractor because he gets to sell five times more welded steel to build individual stalls than he does to build pens. When hundreds of new farms opened up after that, there was a huge shortage of skilled stockpeople, and the contractors touted labor savings and ease of management as a selling point for sow stalls. By then there was an improved electronic feeder on the market, but few people were interested because the early adopters had failed, and of course the building contractors were quick to tell their customers that electronic feeders did not work so they could sell sow stalls.

The building contractors were running the show, and they built what was good for building contractors, not animals. That happens with cows and chickens, too. No company or organization should allow a contractor to dictate design.

Premature Transfer of Technology

One of the most important lessons I have learned in thirty-five years of designing and installing equipment is that transferring new knowledge and technology from the university to industry often takes more work than researching and creating the design in the first place. The field of diffusion research has many examples of good technologies that failed at some stage of the transfer to the market.

Ethologists, veterinarians, and animal scientists need to spend more time transferring the results of their research to industry. Just because you've built a better mousetrap doesn't mean people want a better mousetrap or will pay to have a construction crew install a better mousetrap in their plant. My experience with my center-track conveyor restrainer system for cattle taught me that there are four steps necessary to transfer behavioral research results to business successfully:

1. *Communicate your results outside the research community.* It's important to publish your research in peer-reviewed journals so knowledge doesn't get lost. But just publishing in journals isn't enough. Researchers need to publicize their work by giving talks and lectures, writing articles for industry magazines, and creating and maintaining websites. One of the reasons I was able to transfer cattle-handling designs to the industry is that I wrote over a hundred articles on my work for the livestock industry press. Every job I did, I published an article about it. I also gave talks at cattle producer meetings, and I posted my designs on my website where anyone could download them for free. People are often too reluctant to give information away, I find. I discovered that when I gave out lots of information I got more consulting jobs than I could handle. I gave the designs away free and made a living by charging for custom designs and consulting.

2. *Make sure your early adopters don't fail.* The first people who adopt a new technology have to succeed or the technology may fail. Researchers and developers need to choose companies with management that believes in what they're doing, and they need to stay on top of every detail. I nursemaided my early adopters every step of the way and I made sure everything worked. If I hadn't, I don't think the center-track conveyor restrainer system would be in use today.

3. *Supervise all early adopters to ensure faithful adoption of the design.* After my first plant successfully installed the center-track conveyor, I spent a lot of time on site at the next seven plants to make sure the equipment had been installed correctly. It's a good thing I did because the steel-welding companies were making many terrible modifications on their own say-so. In half

the plants I visited, I found installation mistakes that would have caused the system to fail if I hadn't corrected them.

4. *Don't allow your method or technology to get tied up in patent disputes.* I have seen many sad cases of companies buying patent rights to good technologies to *prevent* a good new design from being adopted. That happened in the pig industry in the 1970s when a designer in Ireland developed a humane, low-cost electric stunner for pigs. He made it out of bicycle parts, which were cheap, and he designed it so it ran automatically and you didn't have to pay an employee to run it. Small companies could afford to buy it, but they never got a chance because one of the big equipment companies that manufactured and sold an expensive stunner bought the patent rights and killed it.

It was a terrible waste because the new design was more humane than the equipment the small plants were using and it probably wouldn't have worked in the larger plants anyway. The company that bought the rights wasn't going to lose any money if the cheaper design went on the market. They just wanted to get rid of anything that competed with their system even hypothetically.

That same company, when I was working on the center-track conveyor system, was working on its own version. After I finished my designs I purposely killed all the patent rights by publishing my drawings in a meat magazine. That put the designs in the public domain. No company anywhere in the world could patent any part of the design. I wanted to make sure it got used.

After I published my drawings I went to a big trade show for the American Meat Institute at McCormick Place in Chicago. The sales rep from the big equipment company was there, and he was so mad he wouldn't talk to me. Today the center-track conveyor restrainer system is used in twenty-five plants in the United States, Canada, and Australia, and half of all cattle in the

United States and Canada are handled in my system when they go to slaughter.

The principles of technology transfer are the same for all species. The pig industry still has a huge need for innovative new systems that will improve pig welfare. When I was in graduate school getting my PhD, Ian Taylor, a student across the hall, was doing a study on the table manners of pigs. On some farms the pigs wasted 10 to 20 percent of their feed and on others there was little feed wastage. Ian figured that the design of the feeder was to blame, so he filmed pigs eating from different types of feeders in slow motion. When he plotted the motion of their heads, he discovered that pigs are real slobs when they eat. When they gobble down their feed, they move their heads all over the place. The only way to prevent a pig from wasting feed is to design a feeder with a large bowl so a pig can be a slob and not push the feed out. This seems obvious, but there are still many feeders with little tiny bowls because his research findings never got fully transferred.

7 Chickens and Other Poultry

I WAS ALWAYS A COW AND PIG PERSON, not a chicken person. I got very attached to cows when I was a teenager and first saw cows being put inside a squeeze chute to get their vaccinations. That's what gave me the idea of building my own squeeze chute to calm down my hyper-aroused autistic system. I was already very interested in animals at the time, and the whole experience motivated me to go into a career designing better handling systems for cattle and pigs. I liked chickens, but I didn't think about working with them, maybe because you can't squeeze a chicken. Being autistic didn't give me the same "in" to chickens that it did to the big animals.

In 1997, I got pulled into working with the chicken industry when McDonald's asked me to help them work on chicken welfare. They were using my audits for cow and pig welfare, and they needed an audit for chicken welfare. The only things I knew about birds came from my college classes and childhood experiences. But that was enough to get started.

Chickens are very social birds that are intensely attached to their mamas. Baby chicks will follow the first moving thing they see after they're hatched the same way geese and turkeys do. In my college classes, my teacher, Dr. Evans, taught us about Konrad Lorenz, the famous ethologist who had a flock of geese that followed him everywhere. He had raised the geese from the day they hatched and they had become imprinted on him. I already knew about imprinting because when I was a child one of the neighborhood kids raised a duckling that followed the family dog everywhere. The duck was imprinted to the dog and lived for many years thinking she was a dog.

When chicks get separated from their mother or the person they are imprinted to, the PANIC emotion kicks in. The chicks' motivation to stay close to their mother is great and chicks will peep, peep furiously when they are alone.

Chickens also have a very active FEAR system because they are a prey species animal. Hens are biologically wired to hide when they are laying eggs. Chickens fly poorly compared to other birds so they can't nest high up in trees, and their wild ancestors, the jungle fowl, got eaten if they laid their eggs out in an opening. All hens want to lay their eggs in a secluded spot where they are concealed but can still watch what is going on around them. This is very important to them.

Jungle fowl had to spend many hours a day foraging to find enough food to sustain them, and chickens housed outdoors do the same thing. They spend most of their day pecking at the ground to find insects and other food. Pecking is instinctual. Baby chicks do not need lessons from their mamas on how to peck at food. I once saw a hen at a petting zoo that was absolutely intent on pecking a rubber band on the ground. I wanted to get it away from her before she ate it, but every time I reached for it, she ran away with it, dropped it, and started pecking at it again. Her SEEKING system was in overdrive. I kept trying but I never did manage to get the rubber band away from her, and she finally ate it.

I didn't know very much about the U.S. chicken industry, either, when I first started working for McDonald's. Of course I knew that animal welfare organizations had a lot of concerns about chicken welfare, but I hadn't been inside a U.S. poultry plant or visited a farm myself, and this was before the days of Internet videos and YouTube. What I saw when I went on my first trips made me go "Aaaaauuuuggghhhh." The welfare of the chickens I saw was horrible.

My first trip was to a broiler farm and a chicken hatchery. (*Broilers* are chickens sold for meat; *layers* are hens that lay the eggs we eat.) I was with executives from McDonald's. We saw very rough handling on the farm by the chicken-catching crew who were loading the birds into the coops for the trip to the meat plant. The workers were picking the chickens up by one wing and flipping them over, snapping the fragile wing bones in two. We even saw workers running live chickens over with a forklift. The McDonald's vice president who was with me said, "This looks like a Humane Society undercover video."

Welfare conditions at the hatchery were terrible, too. We saw a little box filled with half-dead babies, and I asked what they did with them. They told me that the guy who handled sick baby chicks was on vacation and would deal with them the next week when he came back.

That was rubbish.

I said, "Yeah, sure. You're throwing them live in the garbage dumpster and that's going to stop."

The first thing I did was have the hatchery put in CO_2 boxes so they could euthanize damaged chicks. Then I set a limit of no more than 1 percent of the farm's chickens having broken wings during handling and transport to the processing plant.

After that trip, McDonald's asked me to go with them to look at welfare conditions at the farms they were buying their eggs from. On the farm we saw, the birds were stuffed into the battery cages so tightly they were lying on top of each other at night. When we went back to McDonald's corporate office, I took a piece of paper off a printer and I folded it in half and I said, "This is how much space your suppliers give laying hens." Terrible.

After we saw the young laying hens on the farm, I asked to see the old ladies who were at the end of lay. The workers didn't want to do that, so I said, "You can either show them to me or I'm going to walk up and down the row of buildings and look inside all the layer houses until I find them."

They took us to see the almost-spent laying hens, and the chickens were in unbelievably hideous shape, just hideous. The hens were old and had been so hyper they had beaten their feathers off. They were half-bald. They didn't even look like hens anymore.

Chicken welfare is so poor that I can't talk only about the core emotions in this chapter. I have to talk about chickens' physical welfare, too.

Three Problems: Handling, Industry Practices, and Genetics

Physically, chickens suffer for three main reasons: rough handling by workers, bad industry practices, and poor genetics. Genetic problems cross over with the core emotion systems, so I'll start with handling and poor industry practices.

I've described the awful handling I saw, but you can see worse in videos that are posted online. There is one video of slaughter plant workers doing speed handling with chickens, jamming the chickens down in the shackles. The shackles are metal wires that hold the legs of the chickens when they are hung upside down for stunning and slaughter. The workers played stupid games like, "From how far away can I throw a chicken and have it land in the shackles?" There's also another incident in a different plant where the workers were playing "chicken squirts," where they take a live chicken and squeeze it so hard the chicken poops in another worker's face. These are intelligent, sentient, living birds. It's horrible.

Bad handling is a problem that could be solved at the level of individual workers, but the industry has bad practices that would require changes in equipment, too. The worst abuses happen to laying hens during their productive lives and at the end of their lives when they are spent and too old to produce eggs. During their lives they're kept inside tiny battery cages, where the problem isn't just the crowding, but the birds having been so overbred for egg production that their bones are fragile and break easily.

Spent laying hens suffer at the end of their lives, too, mostly because they're not worth anything on the market. They have a horrifically high rate of injury. One survey in England found that 29 percent of battery hens had newly broken bones just before they were stunned.[1] A lot of those injuries happen when the workers pull them out of the too-small doors to the cages and they get their legs or wings caught in the wire frame. There's no financial incentive to handle the birds carefully. If you break a broiler chicken's wing you can't sell it to make Buffalo wings, but if you break a spent laying hen's wing it doesn't matter.

The hens are worth so little that many farms do not bother to send them to a slaughter plant. They kill them at the farm with methods that range from atrocious to barely acceptable. Some of the farms were just throwing the hens, when they were old ladies, into the dumpster alive. Others get rid of their spent hens by sucking them up in a vacuum truck that is used to clean sewers. This sounds like it's just a handling problem, but it's not. When you have hundreds of birds to put down humanely, it's expensive.

On top of all that, a lot of spent laying hens have to be transported long distances to slaughter because not many plants will take them. Instead of getting veterinary care for the injuries they sustained when they were pulled out of their old cages, the hens are stuffed inside new cages and stacked into trucks and driven for miles and miles to the plant. Ian Duncan, a major researcher at the University of Guelph, says, "The disposal of spent laying hens is probably the most serious welfare problem confronting the poultry industry today."[2] I really agree with his statement.

Another major welfare problem is that poultry are put through invasive procedures without anesthetics or painkillers:

- *beak trimming* of laying hens and some flocks of broiler breeder hens, where one-third of the beak is amputated with a heated blade
- *toe dubbing,* where the "toes" of breeder roosters are amputated
- *desnooding,* where the male turkey's *snood*—the long flap of skin that hangs over a turkey's beak—is cut off
- *dubbing* of male chicks that are going to be used as breeders, where the chick's comb is cut off

All of these procedures are painful, but beak trimming is especially bad because beaks have a lot of pain nerves, and some of the hens act like they are in chronic pain after the surgery. They don't peck at feed nearly as much as the other birds.

Beak trimming is done for a good reason, which is to keep the chickens safe from each other. Not too long ago, one of the progressive companies tried to rear a broiler breeder flock of hens without beak trimming after one of their customers asked them to try it. This resulted in a chicken-pecking disaster. The chickens were housed in a really nice cage-free system with private nest boxes and sawdust to scratch in, but the hens still did heavy damage to each other because no genetic selection had been done for good social behavior.

Ways to Improve Poultry Slaughter

The handling methods used by almost all slaughter plants around the world are very stressful. When chickens are unloaded at the slaughter plants, their legs are forced into metal-wire shackles, and the birds are hung upside down by their feet on a shackle line. Some chickens rear up and try to get off the shackles. Then the shackle line carries the upside-down birds along and submerges their heads in a tank of water called the *water bath* where an electric current is run through their brains to knock them unconscious. After that they're moved out of the water bath and their throats are cut.

That's the way it's supposed to work, but you need *very* good handling. When workers don't run the equipment right, a chicken can come out of the bath still conscious and have its throat cut while it's sentient and can feel pain and fear. Handling is especially likely to be bad with the old laying hens because their bones are so fragile that they break when the workers pick them up. Sometimes the birds are still alive and conscious when they are put in the scald tank after slaughter. This occurs in poorly managed plants that are understaffed.

There are solutions for most of these inhumane practices now. A much more humane way to slaughter chickens is gas stunning, which makes the birds go unconscious. It eliminates all the stressful live shackling of the birds. In the best system the birds wouldn't be taken out of their transport cages at all. The cages would just go straight from the truck onto a conveyor belt and then into the gas stunning chamber.

Right now there is a giant fight going on between different researchers on the best gas mixtures to use. Suddenly introducing chickens into CO_2 at too high a concentration causes distress. My view, though, is that some discomfort during gas inhalation may be a small price to pay to eliminate stressful live shackling. Gasping and head shaking may be acceptable, but if the chicken tries to escape from the container, the gas mixture must be changed.

Not too many plants are using gas systems because of another premature transfer of technology from the lab to the industry.

The first gas stunning system a few early adopters installed knocked the chickens out quickly and painlessly, but then a lot of them went into violent seizures and ended up with broken wings that couldn't be used for Buffalo wings. The seizure problem was so bad that the plants couldn't use the gas stunning system on their heavy birds at all. Eventually, the first plant that installed it in the United States ripped the whole thing out and went back to water-bath stunning. That soured the whole U.S. industry on gas stunning, so now any researcher or company trying to get the poultry industry to install a gas stunning system has a major hurdle to overcome.

You cannot take new systems from the lab to industry until all the bugs have been identified and *fixed*. There's a company that has built an improved gas stunning system now. I've told them that they had to have one beta-test plant running at full-scale commercial volume and make it work *before* they try to sell the system to any other plants. Now they have successfully operated the new system in a big chicken plant.

The industry does have two good gas stunning systems for turkeys that work really well. Turkeys weigh over thirty-five pounds and are too heavy for workers to be dragging them out of crates all day long and throwing them into shackles. This is a case where an improvement in welfare for the animals would also be an improvement in welfare for the humans. With gas stunning there's less noise and dust and workers can stand in an ergonomically correct position to do their jobs. Gas stunning would be better for the industry, too, because there's less damage to the chickens and to their meat.

Innovation on the Farm

The problem of doing painful, invasive procedures on chickens without anesthetics or painkillers isn't close to having an answer, but the industry has come up with a solution for the worst of these practices, which is beak trimming. Many large hatcheries have now replaced the hot blade with a device that heats the tip of the chick's beak with an infrared beam. The beam leaves a little brown spot near the tip of the beak, and in a few days the tip of the beak sloughs off. The infrared device is much more humane. I know, because I took part in a comparison of this device with the hot blade. The workers put chicks in my hands immediately after they had been beak-trimmed using the infrared device or the hot blade. I kept my eyes closed so I could not see which treatment the chick had received. It was easy to tell which chicks had been hot-bladed, because their hearts were beating a mile a minute.

The battery-cage situation is better than it was. United Egg Producers (UEP) has issued guidelines requiring that hens have enough space in which to lie down without being on top of each other, and hens now have a space equivalent to three-quarters of a sheet of paper per bird. That is enough to let them lie down and sleep beside each other instead of on top of each other.

Also, new cages are being manufactured without the small doors that make it so hard to pull a chicken out without catching its leg or wing on the wire frame. Instead, the whole front of the cage opens up. I hope everyone will adopt these cages as soon as possible.

Farmers keep the hens in cages mainly because it's more efficient and it keeps the hens clean. The original reason for keeping chickens in cages was to separate them from their feces and keep them free of parasites. However, I have visited really nice cage-free systems where hens are housed in a single layer. The birds have a sawdust-covered floor to scratch in, a roosting area, and private nest boxes for egg laying. The hens stay clean because their perches are built over conveyor belts that carry the poop away from the birds.

The chickens do really well in these systems, but they require a lot more buildings and land compared to cage systems. A number of different cage-free designs are being used in which the birds roost in tiered, stepped aviaries that look like a ten-foot-high sloped wall of chickens. The bird density is much greater and the land and building requirements are much smaller.

Some of these systems don't work as well as single-layer cage-free systems. It's more difficult to handle the manure and keep the environment clean, and there have been more problems with parasites, poor air quality, and high death losses. I have been emphasizing the importance of good management to food animal welfare, but this is one area where some really innovative engineering is needed. I challenge all the clever producers out there to invent new systems.

The one really terrible practice the industry has gone the furthest to eradicate is *forced molting*. In nature, molting takes a long time and happens at different speeds in different birds. Backyard chickens following a natural cycle produce eggs for about a year, taper off, and usually molt in the fall when the days get shorter. During molting, their old feathers fall out and new ones grow in. Almost all hens take a rest from egg laying that begins with molting and lasts anywhere from a few weeks up to several months. After that, they begin laying again, but they never lay as many eggs their second year as they did their first. After the age of two, most backyard laying hens cost more to feed than they produce in eggs.

Natural molting means a lot of lost production time, and the chickens never produce as many eggs afterward. So in the past the U.S. industry used a forced-molting procedure to get a second and sometimes a third year of laying out of hens before sending them to slaughter. Forced molting shortens the time the chickens spend replacing their feathers and gets them back into full production faster. To force-molt an egg-layer flock, farmers shorten the hens' daylight hours to six to eight and starve them for ten to fourteen days. That makes the birds molt and shortens the molting period by eight weeks, but it is very cruel. The hens' mortality rate doubles, they become aggressive, and they develop stereotyped pecking and pacing. They are probably suffering severe frustration of the SEEKING system and overactivation of RAGE. Ian Duncan says, "If any other sentient species were subjected to this degree of food deprivation, it would amount to an offense under most states' cruelty to animal laws."[3] Unfortunately, forced molting by starvation is still common outside the United States, and some people are still doing it here.

Today, researchers have developed artificial molting programs that trigger molting without starvation. Instead, the birds eat a lower-energy feed that has a different balance of minerals from their regular feed. This is good news because if all molting programs were stopped, almost twice as many hens would be needed to produce the same number of eggs, which would double the problem of handling and disposing of spent old hens.

Bad Genes

Chickens have several serious welfare problems that come from bad genetics and can be fixed only with good genetics. The biggest problem in many intensively raised animals is pushing the animal's biology for more and more production. Breeders choose the most productive animals — the fastest growing, the heaviest, the best egg layers, and so on — and selectively breed just those animals. Bad things always happen when an animal is overselected for any single trait. Nature will give you a nasty surprise.

Bone breakage is a very serious problem in both caged and cage-free hens because laying hens have been overselected for egg production. Commercially bred hens put all their calcium and minerals into forming eggshells, and their own bones become depleted. Their bones are so weak that in cage-free systems a hen can break her leg just jumping off her perch. The only way to solve this problem is for the industry to accept the fact that birds with strong bones will produce slightly fewer eggs.

Laying hens have other problems, too, especially feather pecking and cannibalism. *Feather pecking* is what it sounds like: one hen pecks at another hen's feathers or pulls a feather out all the way. Severe feather pecking can lead to *cannibalism,* with the victim hen being wounded and then killed by the hen doing the pecking. Even though a feather-pecking hen can kill her victim, feather pecking probably isn't driven by the RAGE system. We know this because of studies mixing unfamiliar hens together. Aggression goes up, but feather pecking and cannibalism don't.[4] They aren't the same thing.

Instead, feather pecking is probably displaced or redirected SEEKING behavior. It's a kind of foraging or exploration of another bird instead of the ground. We know this because of research showing that chickens housed on litter do a lot less of it. They peck at the litter on the ground, not at each other's feathers. The more active the bird and the more foraging behavior it does naturally, the more likely it is to develop severe feather pecking.[5] Both feather pecking and cannibalism are affected by genetics.

Some modern broiler chickens have genetic problems related to growth. I was shocked to learn at a chicken-breeding seminar that the broiler chicken has been so overselected for rapid growth that its bone physiology is totally abnormal. In normal bone development, the body first "erects" a scaffolding or frame of cartilage and then fills in the frame with minerals that harden into bone. After the bone has hardened, the cartilage dies off through programmed cell death. In broiler chickens, something goes wrong with the cartilage, so the bones don't have support while they're hardening and end up misshapen. I liken it to building a new basement wall and taking down the plywood concrete forms before the concrete has fully hardened. In some of the worst cases, a chicken's feet are rotated almost 90 degrees and the legs are twisted.[6] These chickens are genetically lame. Several studies have shown that lame broilers will choose feed laced with painkiller over their regular feed, and a study of lame turkeys showed that they started moving around a lot more once they were on painkillers. The industry has created chickens that have chronic pain in order to get birds that grow at the far outer limits of what is biologically possible. When an animal's biological system is pushed to the point where the physiology is totally pathological, I get disgusted.

The other problem is that modern broiler chickens have been bred to have stupendous appetites so they'll grow super-fast and reach market weight as soon as possible. The trouble is that the *breeder* chickens, the parents of the broilers, have the same stupendous appetites as their chicks. If you let a broiler breeder chicken eat everything she wants, she will become obese, her fertility will decline, and her life will be shortened.[7] These chickens have to be kept on a strict diet just to maintain normal weight. They act miserable, and many of them develop stereotypies.[8] These birds have low welfare no matter what you do. If you let them eat all they want, they have bad welfare and if you don't let them eat all they want, they also have bad welfare. It's terrible. The industry is going to have to breed parent stock with smaller appetites.[9] There's no other way to fix the problem.

Then there are other genetic problems that no one understands. One of the worst cases was the rapist roosters. I wrote about them in *Animals in Translation.* Fortunately, the broiler industry has made some genetic changes to correct these problems, although there's still a way to go. The rapist roosters violently attack hens and injure and even kill them. Before the 1990s there weren't any rapist roosters. They just suddenly appeared out of the blue. First it was just one strain of roosters that had become aggressive but within a couple of years almost all strains had developed the same behavior.[10] Nobody knows why.

The rapist roosters have two problems: They are hyper-aggressive *and* they have stopped doing the courtship dance the hen needs to see before she will mate. They've lost the little piece of genetic code that makes them do the dance.[11] When the hens don't see the courtship dance, they don't become sexually receptive, which may make the roosters' aggression worse. An unreceptive hen would be a form of frustration because it is a restraint on the rooster's action. So the RAGE system would be activated to some degree.

When I wrote *Animals in Translation* it looked like the rapist roosters were a side effect of the industry's selective breeding program to create chickens with bigger breasts for more white meat. But now researchers aren't sure what caused it, or whether the hyper-aggression and the bad courtship behavior are the same problem or two different problems that happened at the same time. Industry breeding programs are trade secrets. It's obvious the industry is selectively breeding for larger breast size because breast size is getting larger. But we don't know what *other* selective breeding programs the industry might be using.

Ian Duncan has an interesting theory about what might have happened. Dr. Duncan points out that big-breasted male birds have trouble mating because their huge chests get in the way. Male turkeys have such big breasts now that they can't mate at all and the hens have to be artificially inseminated.

Dr. Duncan says that if the same thing is happening to male chickens, the broiler breeder industry may have misdiagnosed the problem. When broiler breeders see chickens with decreased fertility, they attribute the problem to low sex drive. It's possible that the breeders who created rapist roosters were actually trying to increase roosters' sex drive. If the breeders selected for higher libido they could have mistaken a little bit of aggressive behavior toward the hen for higher sex drive and ended up breeding *hyper*-aggressive roosters that for some reason had also lost their courtship dance. We'll probably never know.[12]

Today the aggressive rooster problem has been greatly reduced although it hasn't been eradicated.

Better Breeding Strategies

Most of the time breeders deal with genetic problems by culling chickens that have the problems and mating the ones that don't. Another interesting approach is *group selection*. The researcher Bill Muir at Purdue University has shown that you can reduce feather pecking genetically by using a technique called group selection. With group selection, instead of picking certain individuals, you pick certain *family groups* to breed. Dr. Muir has done this by raising several "sire family" groups — groups of chickens related to each other through their father — and then selectively breeding the *group* that has the highest egg productivity and the lowest amount of feather pecking and cannibalism.[13]

Group selection has a couple of advantages over individual selection. First, the fact that you're working with groups instead of individuals means you know something about the birds' behavior in a group. When breeders select high-productivity, individually housed laying chickens to breed, they don't know whether they are feather peckers or not because they've never lived with other birds. Individually housed laying hens can't express the behavior.

Second, when you have a group of genetically related chickens living together you also see how living in a group affects their behavior and productivity. When you choose which group to breed, you're not just choosing one genetic strain of chickens over another, or two desirable behavioral traits (good egg laying and low feather pecking) over undesirable traits. You're choosing one way of relating to the environment over another.

This is important because most behavior is affected by what's going on in the environment. A hard-wired behavior like the courtship dance is always the same no matter what's going on in the environment. It's like a computer subroutine; once you turn it on it just runs. But everything else is affected by the environment. When breeders use group selection instead of individual selection, they're factoring in the way the selected group relates to its social and physical environment.

Today, only a handful of companies provide all of the commercial layers and broilers around the world, which has greatly narrowed the gene pool. This has created a risky situation because genetically similar animals are vulnerable to the same diseases. Sure enough, when the Australians phased out their homegrown broilers and imported American birds, they ended up with more disease problems.

This is why it's important to preserve the old breeds of animals and poultry. Keeping the classic breeds alive is the only way to preserve genetic diversity and to save animals that have valuable genetic traits breeders may want to breed back into commercial lines in the future. The meat from some of the old breeds is more tender and better quality than meat from animals bred for rapid growth, and the chickens are hardier, too. They perform better in pasture-based or organic farms. They are beautiful, unique animals that shouldn't be destroyed by commercial breeding. Fortunately, many of the older breeds of poultry and livestock are being raised by local farmers and sold in farmer's markets or to gourmet restaurants. If a serious disease ever kills commercial broilers or layers, the entire world will be thanking the small producers and hobbyists who have kept the old breeds of chickens from becoming extinct.

How to Improve Chicken Welfare

The first thing you have to do is raise consciousness. The manager at the egg farm with the almost-bald old lady hens didn't see any problems, and he was upset when I pointed out problems. He responded, "There's nothing wrong with my birds;

they've got good health. I take good care of my birds." He'd gotten so used to seeing those ragged mops that he thought it was normal. This is another instance of *bad becoming normal.* When a welfare situation deteriorates too slowly for workers and management to notice, the new bad situation seems normal. Sometimes it takes an outsider coming in to make people realize that 5 or 6 percent broken wings on broilers or half-bald laying hens are definitely not normal.

After we looked at the chickens we went to the farm conference room, and I said to the company's vice president, "What do you think people from the Chicago airport would say if they could see this?" He'd never thought about it before, so I told him about my ten-people-from-the-airport rule. If I brought ten random people I met at the airport out to this chicken farm, what would they think? Would they say the hens are being housed in decent conditions where they don't experience mental or physical anguish? Or would they say this is cruel and inhumane treatment of innocent animals?

A well-run beef slaughter plant passes the test.

Jam-packed hen cages filled with raggedy, half-bald birds do not pass the test.

Throwing live hens in the garbage does not pass the test.

The vice president of the egg company called me the next day and said, "You made me do some thinking."

To audit the gentleness of chicken handling, I came up with one simple standard: How many broken wings? The big advantage of a broken-wing standard is that I can directly observe broken wings and they cannot be concealed when I visit the plant. Other measures, such as the percentage of birds that die during transport, rely on the plant's records instead of direct observation. Records can be falsified, so I put the major emphasis on things I can see myself such as broken wings, bruises, and breast blisters (when chickens lie in wet bedding, they get ammonia burns on their breasts and legs).

When I first started with chickens, the industry thought having 5 or 6 percent of the birds with broken wings was *normal.* That's completely wrong. A plant should have no more than a 1 percent rate of broken wings. You can't get it all the way down to zero because wing bones are weak and break easily. Perfection is impossible. But plants can easily stay at 1 percent or below.

Lameness scoring systems have been available in the research lab for years, but they were too complex to use on farms. I helped introduce a simple three-point lameness scoring system that measures the outcome of poor breeding, housing, feeding, or management practices on U.S. broiler farms. The test is done on market-ready broilers, and each bird gets one of three scores:

1. Not able to walk ten paces
2. Walks ten paces crooked and lame
3. Walks ten paces normally

When I walk through a flock of broilers, the birds with decent legs quickly move away, which makes picking out the lame or crippled birds easy. On the best farms, 99 percent of the broilers can walk ten or more paces normally. Today the U.S. broiler industry has greatly improved their chickens' legs, but limb deformity and lameness are still big problems in other countries.

Plants have to audit handling. I can't say it often enough. I haven't written an audit for laying hens specifically, but several different audits have been created over the years. I like the one developed by LayWel. LayWel is a research project on the welfare of laying hens that's funded by countries in the EU. Laying hen producer managers should visit the LayWel site, download its new scoring system, and use it to audit the welfare of their birds. The LayWel audit has easy-to-use photos for assessing scores of feather condition, wounds, and bumblefoot lesions on a scale of 1 to 4, and LayWel says it should take one person no longer than thirty seconds to score one bird.[14] Conditions in a chicken house would be assessed by scoring a random sample of one hundred to five hundred birds. Other important things to measure are percentage of soiled birds and ammonia levels.

Economic Factors and Reform

Reforming the chicken industry is tougher than reforming the beef and pork industries for a couple of reasons. One is that there's not very much built-in economic incentive for managers to take good care of their birds. To some degree, it's the opposite. When farmers jam too many hens in a cage, they lower each *individual* bird's productivity, but they get more eggs because they have more chickens. They're financially better off sacrificing high individual productivity for high group productivity. When you sacrifice high individual productivity, you sacrifice welfare.

The other reason why it's harder to reform the poultry industry than the beef or pork industries is that the poultry industry is structured differently. Hamburger is made out of beef trimmings, which are the leftover pieces of meat after the steer or cow has been butchered into steaks, loin cuts, roasts, and so on. To get enough trimmings to make millions of burgers each year, the big restaurant chains buy from lots of different plants. This is true with pork, too. Plants make sausage out of pork trimmings, so they buy from a lot of different plants in order to get enough trimmings. That gives the restaurant chains power over their suppliers. If I have forty beef plants on my supplier list and I take three of them off the approved supplier list, I still have enough plants to buy from.

Chicken is different. Restaurants use only the breast meat for the white meat sandwich. If I'm a big restaurant chain and I'm buying chicken to make chicken sandwiches, I'm purchasing the chicken's equivalent of steak and putting it in my product. I'm not buying trimmings. So I have only four or five plant complexes on my supplier list because I'm using most of the whole bird, and if I suspend one plant for welfare abuses I don't have enough product to sell. McDonald's can throw beef suppliers off their list without affecting the number of hamburgers they serve. They can't do the same with a limited number of poultry suppliers. Most hamburger restaurants don't have the same economic clout over their chicken suppliers as they do over their beef suppliers because most use almost all the production from a small number of poultry plants. They did this because using a smaller number of plants has advantages for the development and launching of specialized custom-made chicken products.

Wendy's is the one chain that has a shot at changing the U.S. chicken industry because they buy chicken from over twenty-seven slaughter plant complexes instead of only four or five because they use standard cuts of chicken. Wendy's can throw a plant off the approved supplier list and still have enough chicken to supply their restaurants. They're doing an excellent job auditing the handling at their suppliers. They've gotten wing breakage down to 0.8 percent in a plant that processes large seven-pound birds. This is a terrific score because it's much easier to break wings on big birds. Wendy's is doing so well that I've changed my audit measurements. Before I got the new scores from Wendy's I had no more than 1 percent wing breaking for light birds and no more than 3 percent for heavy. Now I'm changing it to no more than 2 percent for heavy birds because Wendy's has the data to show it can be done.

Wendy's has gotten lameness scores way down on the farms, too. Ninety-nine percent or more of the eight-pound jumbo birds can pass the ten-step chicken lameness test on their best farms. When I first started working with broiler chickens, even the lightweight birds had high percentages of twisted legs and lameness.

Things are finally moving in the right direction in the poultry industry, although there are some top executives in the U.S. poultry industry who don't get it yet. Some of the complex managers have been absolutely wonderful to work with. They have quietly worked on being progressive even though they receive little support from the head office. Chicken welfare is improving in other countries, too. In 2006 I went to Australia and saw that they were doing really progressive work with broilers to correct leg problems caused by rapid growth.

Unfortunately, even when you combine Wendy's twenty-seven plants with the plants supplying Burger King and McDonald's, which also audit their suppliers for welfare, you're still auditing only 30 percent of the poultry complexes compared to 90 percent of the beef industry. That's not enough. The other 70 percent of the plants sell to supermarkets that either do not audit or have auditing programs that are less strict.

The handling isn't bad at all of the unaudited plants. About 20 percent of these plant complexes handle their chickens gently and do not need to be audited. Everything depends on the manager. There have always been good managers who care about the animals and manage their workers well. Unfortunately, poor management means poor welfare. When you see videos of employees abusing chickens, that's a management problem. There's a famous video of workers throwing chickens and stomping on them at a Pilgrim's Pride plant. Good managers don't let that kind of stuff go on. The management at that plant has now installed video cameras to monitor the workers. Video cameras really help to prevent bad employee behavior. The best manager I ever saw was in a beef slaughter plant. He had his office in a room overlooking the cattle-handling facilities so he could make sure no bad stuff was going on. Good managers in poultry plants keep a close eye on the employees and their handling of the birds.

Besides doing chicken welfare audits, large meat buyers can also provide incentives to their suppliers to improve systems. For instance, Burger King, Safeway, and other companies have agreed to buy a certain percentage of their eggs from cage-free hens — although these systems need to be monitored, too. We can't just assume that cage-free means high welfare. I have seen excellent cage-free systems and I have seen filthy, horrid, crappy ones.

I've also seen an interesting provision in the contract one major chicken company uses with the family farms that raise their chickens. The person signing the contract has to be a primary caretaker of the birds. The farms can't just turn their chickens over to hired help. This reminds me of a talk I heard Earl Butz give about collective farms in the old Soviet Union. Earl Butz was secretary of agriculture under Ronald Reagan. He told us, Do you think that the person taking care of the pigs is going to stay up all night watching the Soviet's sow have the Soviet's pigs? The answer is no.

This is true all through business. I've had building contractors tell me that when you assign a particular truck to a worker instead of rotating the fleet, he'll take care of it because he thinks of the truck as his. That's why the company I mentioned requires that the person signing the contract be a primary caretaker. He's the one who will get extra money for taking good care of the chickens.

Farms and Slaughter Plants Should Have Glass Walls

My last recommendation is that farms and slaughter plants should have glass walls. I tell executives, "There's this wonderful technology you can use to improve animal welfare. It's called glass. It's called webcam." People need to see what's happening on farms and inside plants. Michael Pollan, in his book *The Omnivore's Dilemma,* has also stated that slaughter plants should have glass walls so that people can see what happens inside.

Poultry industry executives always object to audits for reasons of *biosecurity,* which is a legitimate concern. An auditor could walk through a plant, step in some chicken poop, and spread an avian flu virus to the next farm. Webcams can solve this problem. Install cameras in the plants and farms and let everyone see what's going on.

Transparency has a powerful psychological effect because people and animals behave differently when they know someone is watching. Some professors in the psychology department at Newcastle University did an interesting experiment on this. The department had a coffee station that ran on the honor system. People were supposed to fix their coffee or tea and leave a payment. However, a lot of people were just helping themselves and not leaving any money. So three of the professors ran an experiment to see if they could change people's behavior. One week they put up a poster of flowers at the coffee station, and the next week they put up a poster of a pair of eyes looking at the person getting coffee. Then they switched back again. They found that people were 2.76 times more likely to pay for their coffee or tea when the poster with the eyes was there.[15] Just the thought of being watched improved people's behavior. That's why we need webcams inside slaughter plants, hatcheries, and farms. People behave better when outsiders can see what they're doing.

When I made it clear that webcams were a good idea, some people in the industry went berserk. They thought I was off my rocker, and I was shocked at how vehemently opposed they were to the idea. Thirty-five years ago, when I first started working with cattle, schools went on field trips to a local slaughter plant in Tucson, Arizona. Back then a slaughter plant was something normal, not a mysterious place hidden behind a wall. Today there is a wonderful small organic slaughter plant in Minnesota that has a viewing area. People can come and watch the entire process. Most of the visitor reactions have been favorable. Another great alternative is auditing by an auditing company over an Internet connection. This is already being done in several innovative plants.

Blue-Ribbon Chicken Emotions

Cramming hens into tiny cages is a form of restraint, and restraint activates the RAGE system in the brain. We've got to find a way to give laying hens more room. Broiler chickens living in large groups on a sawdust-covered floor have much more freedom of action. Their RAGE systems probably aren't being overactivated.

Both laying hens and broiler chickens experience a great deal of FEAR at the end of their lives, when they are forced into tiny cages for transport to the slaughter plant, and then dragged out of the cages, jammed into shackles, and carried upside down to the water stunning bath. Chicken slaughter has to be changed. No animal should spend its last conscious moments in a state of terror.

Laying hens probably also suffer a lot of FEAR throughout their lives. Chickens are prey animals, and they have an instinct to hide while they're nesting. Wire battery cages stuffed with four or five hens make chickens feel vulnerable. The hens can't nest by themselves and they can't hide, either. In Europe they have built small plastic nest boxes with a plastic flap on the front, like a soft doggie door, inside the cages. The chicken can hide inside the box, and she can poke her head out of the flap to see what's going on. Laying hens need nest boxes to have decent emotional welfare.

The Europeans have installed *furnished cages* in some large commercial systems, which will provide more enrichment than a standard battery cage. One of the most important items that would be placed in the furnished cage is a nest box. These cages are also taller and provide enough space that a hen can stand at full upright position and fully extend her wings. Furnished cages would reduce both the RAGE that comes from being restrained and the FEAR that comes from being exposed to potential predators.

The other two items people want to put inside furnished cages are a substrate where the birds can take a dust bath and perches where they can roost. Both of these behaviors are hard-wired, although there is an element of learning involved in roosting. Adult birds that never had perches when they were young have difficulty using them. The question is: Do chickens *need* to do natural, hard-wired behaviors in order to have good welfare? Or can they live happily without some of these behaviors?

This gets back to the question of which is more important, the behavior or the emotion that drives the behavior. Perching may be associated with PANIC or FEAR or both because subordinate birds can get up on perches and stay out of harm's way. One study found that subordinate birds on perches don't get feather-pecked as much.[16] (Perches are probably good for chickens' leg bones, which are a reason to provide them.[17])

Dust baths are probably driven by the SEEKING emotion.

Dr. Duncan did an experiment using weighted doggie doors to test chickens' wants and needs. The question was: How much weight will a chicken lift in order to get something she wants? The heavier the doggie door the chicken was willing to lift, the more she wanted the thing behind the door. He found that a chicken really wants a secluded place to lay her eggs. A hen's motivation to get to a nest box was equal to her motivation to get to a food tray after thirty hours off feed.

A dust bath, on the other hand, is probably a luxury.[18] A hen wouldn't lift super-heavy doggie doors to get to a dust bath. Her attitude is, "Yeah, I would have liked to get to that dessert, but I can do without it." That's a good thing, because the poultry industry would have a very hard time putting dust or anything resembling dust inside a battery cage. Dust would wreck the conveyor belts that are used for collecting the eggs. About the most the industry can do without destroying the belts would be to add a solid floor to part of the cages and let the hens scratch a bit of feed. Ian Duncan didn't include perching in his study, so we don't know how strongly motivated chickens are to get up on a perch.

Grown chickens have problems with aggression, so farmers have to keep that in mind. Stressed chickens get aggressive with each other, especially when they're mixed with unfamiliar birds. (Some researchers think that the practice of dubbing — cutting off the male chick's comb — may make it hard for chickens to recognize each other, because chickens use the head region to tell each other apart. No one knows for sure, but if that's true then dubbing may increase aggression.) Fortunately on most farms, chickens live their entire lives with the same group of birds.

Farmers also need to avoid mixing physically active chickens with more passive chickens because inactive birds tend to get picked on. One study found that when feather-pecking started up, birds that were standing, sitting, or lying still were more likely to get severely pecked than birds that were dust bathing, rubbing their heads, rubbing their sides, shaking their wings, or lying on their sides and scratching themselves.[19] Similar behavior occurs in pigs. When Bill Muir used group selection at a commercial pig company to produce a line of gentle, non-aggressive pigs, they could not be mixed with the more aggressive genetic lines. The gentle pigs got completely beat up.

Active hens shouldn't be caged with inactive hens, and if chickens are kept in pens, the pen should be structured so that it has an obvious resting place for the chickens that's separate from where they eat. That keeps temporarily inactive chickens away from chickens that are in food-SEEKING mode.

Another good modification to the environment is to build partitions and places to hide inside each pen. When one animal gets attacked by another animal, it needs to have a place to get away to.

There is one other welfare issue for chickens, which ethologists have different views on. Baby chicks are hatched in an incubator and never see their moms at all. This is very unnatural since, in the wild, baby chicks imprint on their mama. The reason ethologists have differing opinions on this practice is that chickens are *precocial*, which means that they're already pretty well developed when they come out of the egg and can look out for themselves. The baby chicks are never alone; they're always with other baby chicks. It's possible that's enough to keep the PANIC system turned off. It's definitely important for chickens to be with other chickens. In nature, a single bird is never isolated. A solitary chick would experience both FEAR and PANIC.

I've saved SEEKING for last. Chickens are smart, fast, not very complicated, and cautious, which tells me we might be able to do more for chickens using simple, inexpensive enrichments than we can for any other captive animal. We have some evidence this might be true. One study found that adding a simple string device — strands of white string the chickens could peck at — lowered feather pecking in White Leghorn layers, which are genetically prone to high levels of pecking. What's interesting is that the chickens in the study didn't get tired of the string device the way pigs get tired of old straw after just a few hours. After fifty-two days, the same pieces of string were still interesting to the chickens. In one experiment they pecked at the string just as many times on the first day as they did on the fifty-seventh and fifty-eighth days.[20] Chickens may not have as strong a need for novelty as other animals.

If that's true, it's all the more reason for the industry to give chickens simple enrichments like string devices. A little goes a long way with a chicken. Laying hens have the poorest welfare of all the farm animals. If we can make their lives better by giving them simple pleasures inside their cages and pens, we have to do it.

8 Wildlife

I AM WORRIED about whether we will always have a Jane Goodall.

Jane Goodall wasn't a trained ethologist. She went to secretarial school and then saved up money to visit a friend in Kenya. While she was there, she met Dr. Louis Leakey, the famous anthropologist, who hired her to be a secretary in his museum and eventually set her up to study the chimpanzees at Lake Tanganyika. Dr. Leakey also sponsored Dian Fossey to study mountain gorillas in Rwanda. Dian Fossey's degree wasn't in ethology, either. She had a degree in occupational therapy.

Jane Goodall eventually did get a PhD in ethology, but not until after she made two major discoveries about chimpanzees: She discovered that they ate meat and used tools, at a time when scientists believed that the fundamental distinction between humans and animals was that humans used tools and animals didn't. When Jane Goodall reported that she had seen chimpanzees using twigs to fish for termites in termite nests, Dr. Leakey sent her a cable that said, "Now we must redefine 'tool,' redefine 'man' or accept chimpanzees as humans." She also saw the chimps eating meat, another thing everyone knew chimpanzees never did.[1,2]

Jane Goodall went in the back door to become an ethologist.

That's something I've thought about a lot, because people with autism usually have to go in the back door. We have a lot of trouble following the normal paths. We don't do very well in interviews, which is a big problem for us, and a lot of autistic people also have extremely "uneven" academic skills. An autistic person can test at the bottom of the IQ range on one subscale and at the very top on another. For example, I had great difficulty with algebra because there's no way to visualize it.

I couldn't be doing what I'm doing if there weren't any back doors. When I was working for *Arizona Farmer-Ranchman* magazine I almost lost my job after the magazine was sold and a new boss came in who thought I was weird. The nice graphics lady, Susan, saved me. Susan told me, "He's going to fire you. We need to make a portfolio of your work." We got all of my articles together and gave them to Jim, and instead of firing me he gave me a raise. That's a case of going in the back door, not to get a job but to hold on to a job. I didn't have the social skills to detect that my conversation had made a bad impression on Jim. Susan told me I had to sell myself to my new boss by showing him the high quality of my work.

My dissertation adviser at the University of Illinois, Stan Curtis, held open the back door for me, too. My problem was math. At Arizona State, where I got my master's degree, I got through my stats course with a lot of help from my tutor, another graduate student named Raquel. Raquel's mother owned a bar in downtown Phoenix on skid row, and Raquel said my payment would be for me to repair the rotted-out floor in the bar. The floor had decayed because the beer drain filled a fetid swamp underneath the floor. It was worth it, because I ended up with a C in the class and I got my master's degree.

I wasn't as lucky at the University of Illinois, where I went for my doctorate. The statistics class they required for the degree was terrible. I had a tutor there, too, who had to teach me everything from the bottom up. Instead of teaching me the formula for chi-square tests first and then giving examples, the way my professor did, she gave me examples of specific experiments that would use chi-square and *then* give me the chi-square formula. I made up a book with fully worked-out examples of experiments I would be likely to do as an animal science researcher.

I still couldn't pass the tests in class, even though I knew how to use statistics for individual experiments, so my tutor showed the department all my work. Dr. Curtis knew I had passed my course at Arizona State with a C, and he could see that I knew what tests to run in different kinds of experiments. Also, I was doing well in all my other classes. So he did not make me take statistics again, and I got my PhD.

There are probably a lot of people who can flub interviews and classroom tests but still be good fieldworkers. Some of the best fieldworkers I know have Asperger's syndrome or they are dyslexic. But I think it's getting harder to do things without going through the proper channels. How many people would even try to be Jane Goodall today? Jane Goodall was a superb fieldworker who lived with animals, observed them closely, and understood them. She did her work in the field, not behind a computer making mathematical models of chimpanzee populations.

But good fieldworkers sometimes come through more conventional academic channels. Jill Pruetz, the researcher who discovered that chimpanzees use weapons to hunt prey, is a professor of anthropology at Iowa State University. She didn't go to secretarial school. Dr. Pruetz has spent seven years living in a twenty-three-square-mile area in Senegal so she can study the savanna chimpanzees that live there. She spends thirteen to fifteen hours in the field, six days a week. The chimps' range is so big that she has to get to the chimps before they wake up and then follow them all day long until they go to sleep at night. Otherwise, she might not be able to find them again. At night she stays in a Fongoli village with thirty villagers and one latrine. She's been ill with malaria seven times.

She spent four years just habituating the chimpanzees to her presence before she could study them. Then she spent three summers observing their lives. She discovered that some of the chimpanzees make spears out of tree branches and use them to spear bush babies inside hollow trees.[3,4] Bush babies are small furry animals. The chimpanzee breaks a branch off the tree, strips off the leaves, and sharpens one end to a point with its teeth. Then it stabs the spear violently inside the hollowed trunk to kill any bush baby that might be inside. This discovery is so revolutionary that it has caused a big controversy in the field of primate research, because it is the first documentation of an animal using a tool as a weapon for hunting.

Too Much Abstractification

Dr. Pruetz is one of the few young professors doing intensive fieldwork in animal behavior. Animal research is getting more and more what I call "abstractified." Instead of people studying the real animals in their natural habitats, researchers use fancy statistical software to construct statistical models, and then they study the models. One of the reasons this has occurred is the development of new mathematical tools. Many academic scientists think that unless you use sophisticated mathematics, your study lacks scientific rigor. If you look at one of the main textbooks on wildlife management, *Wildlife Ecology, Conservation and Management,* second edition, by Anthony R. E. Sinclair, John M. Fryxell, and Graeme Caughley, practically every page is about mathematical modeling: geometric population growth, exponential models, response curves of predators at different prey densities, the Ricker logistic model, the theta-logistic model, and so on. You can't even read the formulas in the book unless you know calculus.

It's the same in the food industry. Mathematical models and statistics are good things, but you have to know their limitations. I can remember sitting through tons and tons of presentations on mathematical models at agricultural engineering conferences in the 1980s, and never once did anyone give a paper where they had taken all their mathematical formulas and actually verified them in the field. I'd raise my hand and go, "Does it work? Did you ever try this out in the feedlot?" No, they hadn't.

I remember one time some university researchers constructed a scale model made out of plywood to test a water-jet system they had designed to blast the muck out of the alleyways between the dairy cattle pens. They put real cow dung inside the model and tried to flush it out. The problem was that cow poop has pieces of straw in it, and when you put cow poop inside a small plywood model, those little pieces of straw are proportionately the size of fireplace logs in a real alleyway. The researchers didn't think about the fact that the water is going to flow differently in a little six-inch alley with straw inside from the way it does in the real world. The model didn't scale up; it didn't behave the same way in reality as it did in the model. That's abstractification.

Without fieldwork, no one would have discovered that chimpanzees use weapons because no chimpanzee in captivity is going to need to make a weapon in order to kill live prey and eat it. I doubt anyone would have seen chimpanzees using any kind of tool.

Thanks to fieldworkers we also know much more about dolphins and whales than we did years ago. Researchers have learned that dolphins live in large, complex social groups in which individual dolphins form and switch alliances with other dolphins. Fieldworkers have even observed "alliances of alliances," where one group of allies forms an alliance with another group of allies. These social groupings may be similar to the cliques people form. Dolphins and whales are multicultural, too. Different groups have different dialects and feeding and play behaviors.[5]

Dolphin play is an especially interesting area where good observational research is being done. There have been a lot of reports that dolphins intentionally blow bubble rings, which are large circles of water similar to smoke rings, to create their own toys. People assumed that dolphins blow bubble rings on purpose partly because their breathing is under full voluntary control. Dolphins and whales are the only mammals like this; breathing is automatic in all land mammals. From what people could see of the bubble rings, it looked like it probably takes practice to blow a good one, the same way it takes practice to blow a good smoke ring.

Researchers studied dolphins that made six different kinds of bubble rings and had at least three different ways of playing with them. One thing the dolphins liked to do was smash the bubble rings. Other times they would produce a vortex around the ring that flipped it 90 or 180 degrees, or they would aim a second bubble ring at the first ring, which sometimes caused a third ring to form. The researchers also reported that a baby dolphin living separately from the four juveniles in the study watched them when they made bubble rings, and then practiced making its own bubble rings.[6]

I think all researchers should do fieldwork, or else work closely with a collaborator who does. Even though I'm not good at math in the abstract, I am good at troubleshooting statistics when something doesn't come out right in a study because I have lots of experience with the real animals. I've had graduate students who get nothing out of their statistics, so I look at the data and say, "Let's put this variable against that variable and see what happens." I do creative sorting; I see things in the numbers that other people don't because I'm not looking at numbers; I'm looking at animals and visualizing them when I sort variables on a spreadsheet. I never get away from the animals.

Good fieldwork is *observational science.* Some people think if you don't have a control group it's not science. But there's a lot of research where there's no way to create a control group. I used to argue with my adviser about this at the University of Illinois. What is astronomy? It is observation. You look at things.

For a lot of research you have to start with observation before you can create a lab experiment or do a statistical analysis. A good example: Epidemiology always starts with observation. People notice that smokers seem to get more lung cancer than nonsmokers, and then an epidemiologist uses scientific data collection and statistical tests to find out if there is a connection between smoking and lung cancer. Research on animal behavior should start with observation as often as possible. The Japanese researcher Tetsuro Matsuzawa is the only primatologist I know of who does both fieldwork and lab research. He has been in the news a lot for his studies of chimpanzees beating humans in a memory task. In the experiment the chimpanzees and humans have to remember the position of a sequence of nine numbers on a computer screen. The chimpanzees always beat the humans. This is another revolutionary finding.

Scientists have been doing laboratory studies on chimpanzees for years, but those studies took it for granted that chimpanzees had "worse" cognitive capacities than humans. I don't know of anyone who has done an experiment to find out whether chimpanzees might have cognitive capacities that are *better* than humans' cognitive capacities. I think a big part of the reason no one ever asked that question is that the researchers were all lab researchers. They didn't study chimpanzees living in their native habitat where they might look smarter than they do living in a pen.[7, 8]

Since Dr. Matsuzawa was studying chimpanzees in the field, he noticed that chimpanzees can do a lot of impressive things. One was that they can tell the difference between two hundred different kinds of plants in the jungle. Dr. Matsuzawa says chimpanzees have a botanist's memory for plants. They remember everything: the individual plants, the time of year they grow, their locations, and their uses. No one working with chimpanzees in a lab setting is going to notice that. Fieldwork helps animal scientists ask the right questions.

Another researcher, Nicola Clayton, who studies birds, also combines lab research with observation of bird behavior outside the lab. She is a biologist who started out researching the hippocampus in Eurasian jays and other birds that hide their food. Dr. Clayton found that the more food the birds hid, the bigger their hippocampus became. Traditionally, that's where lab research would stop, with the researcher just looking at an isolated part of the animal's brain or body. But Dr. Clayton paid attention to what birds did in the wild, too. When she moved to the University of California–Davis, she noticed that western scrub jays were stealing pieces of people's food, hiding them, and then coming back later, not to eat the food but to move it to a new hiding place. Why would they do this? She wanted to know.

That was the jumping-off point for her lab research. She set up an experiment where she put scrub jays in a kind of "suite" with three miniature rooms, like a hotel suite. The birds slept in the center room. Then the next day the researchers moved them to one of the adjoining rooms. One room had lots of food; the room on the other side had nothing to eat. After a few days, the birds could choose which room they wanted to go into, and Dr. Clayton put food in the center room. When the birds found the food in the center room, they took it into the no-food room and hid it in an ice cube tray filled with sand that Dr. Clayton had put there. Dr. Clayton says this is evidence that the birds were planning ahead for the next time the experimenters put them inside the no-food room.[9]

That is a major discovery. The only animal that we know for sure plans ahead is humans. There've been a few studies saying that some primates can do it, but all of those animals were given lots of chances to learn to plan ahead, so they might be showing the effects of training, not of their own planning.

In some of her other research, Dr. Clayton found that jays steal each other's food, and that they remember where they've hidden food that decays quickly, such as worms, versus where they've hidden seeds and nuts. Dr. Clayton interprets this to mean that birds have episodic memory, which is memory for particular events or episodes in the past. That is another huge finding that wouldn't have come about if the researcher hadn't observed the way jays acted in the wild.

Recently, Dr. Clayton has been working with a learning theorist at Cambridge named Anthony Dickinson. Dr. Dickinson says that when he first heard Dr. Clayton say birds had episodic memory, "it was an outrageous statement." I do not think the statement that birds have episodic memory is outrageous. I think it is right. In *Animals in Translation*, I wrote about animals having autistic savant skills. Several scientists have challenged us and do not believe it.[10,11] They say that it can't be true because people with autism are damaged and animals are normal.

This is irrelevant. The memory feats of food-caching animals that can remember the location of hundreds of food stores are highly similar to the ability of some people with autism to memorize every street in a city. My theory is that savant-type skills occur when memories are sensory-based instead of language-based. Language leads to abstractification and loss of detail. Animals naturally lack language and autistic people have language problems because of a disorder, but in autistic people and animals the cause of sensory-based memory is the same: thinking and remembering in pictures instead of words. It is definitely possible to have episodic memory in pictures instead of words. I have many visual memories of specific events.

Fieldwork Outside the Field

"The field" isn't just the jungle or Yellowstone Park or the Serengeti Plains. The field can be any place you find animals, including farms, ranches, feedlots, and research labs. As long as you're doing *naturalistic observation* of the animals, you're doing fieldwork. Naturalistic observation means the scientist observes a naturally occurring situation without trying to manipulate what happens. Naturalistic observation in a research lab would mean that the scientists observe what the research animals are doing when they're *not* in a formal experiment.

There's a book out on Nim Chimpsky, one of the two chimpanzees that were raised with a human family so researchers could see whether they would develop grammatical language.[12] Reading the book, I noticed that the most interesting things Nim did, didn't happen during the formal experiments. They happened in between the experiments, when he wanted something. I became totally convinced that Nim understood sign language when he banged on a closed door and signed "hurry open now."

In her book *Don't Shoot the Dog!* Karen Pryor says animal trainers can learn a lot about an animal's mind and emotions by observing how the animal reacts to *reinforcement*.[13] She points out that if she sees a dolphin in a group of other dolphins jump up in the air and come down with a big splash, she doesn't know why he did that. But if she forgets to give a dolphin she's training a fish when he knows he's supposed to get a reward and he jumps up in the air and comes down with a huge splash directed right at her, then she knows that at least some of the time *jump-splashes* are "aggressive displays."[14] Different animals react different ways to trainers forgetting to give them the reinforcement they've earned, and those differences tell you something about the nature of their different personalities. You get that information only from close observation. Good science depends on good observation no matter where you are, whether you're inside the lab or outside.

Protecting Wildlife through Fieldwork

You have to have good fieldwork to protect wildlife. Cheetahs are an example. Laurie Marker, an expert on cheetahs who studies them in the wild, says that there are only 12,500 cheetahs in twenty-six different countries, which is the lowest population in 9,000 years.[15] What's especially dangerous is that all cheetahs are so genetically similar to each other that they're almost clones. Geneticists say that would have happened because cheetahs went through a population bottleneck around 120,000 years ago when a catastrophic event — probably the Ice Age — wiped out almost all the cheetahs living at the time. The cheetahs that survived had to interbreed to reproduce. As a result, the 12,500 cheetahs that are alive now don't have enough genetic diversity to protect them from the next crisis. If one cheetah dies from the next bad feline virus that comes along, there's a chance that all the cheetahs could die from it.

Cheetahs are going to need a lot of human intervention to survive. The current strategy is to protect cheetahs in the wild while breeding them in captivity as an insurance policy. Cheetahs have been especially hard to breed in captivity, but fieldwork has started to make it easier. Tim Caro's field study of cheetahs on the Serengeti Plains was a breakthrough because he found that male and female cheetahs don't live together in the wild. Zoos realized from his work that they needed to house their males and females in separate quarters.[16]

Laurie Marker made the other big discovery, which was that in cheetahs the female does the choosing. Normally when you're trying to get animals to breed you might bring a male several females he can date and court. He gets a choice. That doesn't work with cheetahs. In cheetah society it's ladies' pick. You need to give her different boys she can choose from, and she's very picky. Even worse, Laurie Marker also found that when you put female cheetahs together, they suppress each other's sex hormones, so if you want to breed them, you must house them separately. Thanks to all this good fieldwork, zoos are having better success breeding cheetahs in captivity.

It's going to take good fieldwork to solve the terrible problems elephants are having.[17] In Africa, elephants have been attacking and killing human beings and destroying villages and crops. Male elephants are raping and killing rhinoceroses and other elephants. The situation is so bad that in 2005 there was a report in *Nature* called "Elephant Breakdown."[18] One of the authors is the research director at the Amboseli Elephant Research Project in Kenya.

The authors of the paper believe that elephants have experienced so much trauma that they have developed PTSD. They show all the signs of PTSD: abnormal startle response, depression, unpredictable behavior, and hyper-aggression. A lot of the young males especially have become violent and are killing humans for revenge. The trauma comes from young elephants watching the matriarchs of their families get killed by poachers. Their elephant culture, which could have helped them deal with their trauma without becoming violent, has been destroyed. It's very bad to kill the older elephants. It's the older male and female elephants that keep the young males in line.

There is no way to research this situation in the lab and find a solution. It's going to take close observation of elephants and hands-on rehabilitative work rearing orphaned elephants. The David Sheldrick Wildlife Trust is one organization that's trying to find a solution. They take little tiny baby orphan elephants that aren't wrecked yet and raise them not to have disgusting behaviors. For the first two years, people hand-feed the babies. During that time the baby elephants form an intense attachment to their keepers. When the babies turn two, the workers mix them with other young elephants ages three to five. At this stage the two-year-olds are still getting milk from people.

Gradually, the older females in the group start to become natural matriarchs. While this is happening the workers slowly wean the young elephants away from their human keepers. They have to do it slowly because the babies are in love with their keepers. Baby elephants have to be taught manners by older elephants, so the two-year-olds stay with the group during the day and come back to sleep with their keepers at night. The keepers teach them some manners, too; even when the babies are really young they're never allowed to physically push a person.

Mountain Lion Attacks on People

Africa isn't the only place having dangerous problems between people and wildlife. We need good fieldwork in this country to keep people from getting attacked and killed by predator species such as mountain lions and grizzly bears. We also need educated citizens who know enough to listen to fieldworkers.

The Beast in the Garden, by David Baron, is a really good book on this subject. Mr. Baron starts the book with a search for a missing jogger outside a small town close to Boulder, Colorado. When the searchers find the jogger's body, at first they think he's been attacked by a crazed serial killer:

> The body, clothed in athletic gear, wasn't sloppily mangled; it was carefully carved, hollowed out like a pumpkin. Someone had cut a circle from the front of the sweatshirt and the turquoise T-shirt beneath, sliced through the skin and bones, exposed the chest cavity, and plucked out the organs. After conducting this ghoulish backwoods surgery, the killer had removed his victim's face and then sprinkled moss and twigs on the lower torso as if . . . performing a macabre ritual.[19]

While the men are standing around staring at the body, one of the searchers looks behind them and sees a mountain lion sitting on its haunches, watching them. The lion was the killer.

Everyone had always thought mountain lions don't kill people, and at that time, in 1991, no mountain lion had killed and eaten a human being in over a century. But some of the fieldworkers who studied the animals knew that their behavior was changing. Mountain lions had always stayed away from humans, but now they were coming close enough to people's houses that residents could watch them out their windows. The fieldworkers knew that was a very bad sign, but residents and even the professionals in the Wildlife Department didn't see anything wrong with it. Everyone liked living close to nature and they interpreted the lions' behavior as a sign that humans and wild animals could live together peacefully.

But that wasn't what was going on inside the lions' minds. What was going on inside the lions' minds was that they were learning that people could be prey. The mountain lions weren't getting tame; they were gradually developing stalking behavior. First they came closer and closer to people's houses, then they picked off some pets, and finally they started attacking people. But no one would listen when fieldworkers told them that's what was happening. Even after the jogger was killed, citizens still didn't want to do anything about the lions.

The other really bad thing that happened, and that's still going on today, is that people are acting like prey. Humans are jogging and riding bikes in the woods. Fast motion is prey behavior and it triggers the prey chase drive in predators. Nobody should jog or bike in an area that has predator animals living in it — and nobody would if people listened to fieldworkers who know how wild animals think and act. When I was in my teens in the 1960s, I went to boarding school in New England, and we often hiked in the woods even though everybody talked about George the mountain lion that lived in the area. One night, George was seen crossing the parking lot, and nobody worried about him. I made a "George" costume and wore it for winter carnival. In the 1960s people did not jog or bike in the woods, so George never bothered anyone.

Preserving the Ecosystem

You need fieldwork to preserve the ecosystem animals live in, too. *Scientific American* had an interesting article about bears and forests. Bears eat salmon, and in the 1940s Alaskan fishermen were so worried about bears eating all the salmon that they wanted to have a big culling operation. That didn't happen, and it's a good thing it didn't because two field researchers named Scott Gende and Thomas Quinn have discovered that if you don't have bears to eat the salmon you might not have a forest, either.

The way it works is that bears kill lots more salmon than they eat, and they eat only the good parts of the salmon. Sometimes they just eat the eggs and throw away the whole fish. Also, because bears are mostly solitary, they don't eat the salmon right there in the water where they caught it. They take it into the forest so they won't get in fights with other bears trying to take their food away. When they get to someplace safe, they eat the part of the fish they want and then leave the rest of the body behind.

That's good for insects, birds, and smaller animals that eat the carcass. It's also good for the forest because once the other animals have finished with the carcass, it breaks down and fertilizes the soil. These innovative researchers found that in some places 70 percent of the nitrogen in streamside trees and bushes comes from salmon. Scott Gende and Thomas Quinn say, "The bears are truly ecosystem engineers: they deliver marine-derived nutrients to the riparian [streamside] system."[20]

Grazing is another example where fieldwork has shown that people's assumptions about what's good and bad for the environment were wrong. There are people who say livestock are bad because they wreck the range. A lot of times that's true. But when cattle or other grazing animals are used correctly, they *improve* the range. As a matter of fact, you have to have grazing animals in order to have decent grasslands at all. The Great Plains were created by the bison.

We learned this from fieldworkers who studied the big migratory herds of wildebeests that used to roam the plains of Africa. Grazing animals in a wild herd always stay close together to give each other maximum protection from predators. Wildebeests have four predators — lions, cheetahs, hyenas, and hunting dogs

— so they were always tightly stocked. They had to eat everything on the plain, not just the good stuff. They would come into one part of the pasture and mow it down, but then move to a new patch of ground long before they damaged the pasture. While they were eating they were pooping on the ground and fertilizing the plants. When they finished grazing and fertilizing one area, they moved to another area, and the grass and forage in the grazed areas had plenty of time to regrow before they returned.

Grasslands can't exist without big migratory bison herds or properly grazed domestic cattle to mow it. Allan Savory, an expert on the ecology of land development, says there are two kinds of land in the world: *brittle* and *nonbrittle*.[21] Brittle land easily turns into desert; nonbrittle land doesn't. The major difference between the two is that nonbrittle land has moisture year-round whereas brittle land has dry spells. The difference isn't total rainfall. There are parts of Africa with high rainfall where the land is still brittle because those areas have one long dry spell each year. Some places go as long as eight months without any rain and then get sixty inches of rainfall during the other four months.

Allan Savory discovered what happens when you let these lands lie fallow — when you remove all grazing animals. Nonbrittle areas revert to forest, but the brittle lands turn into desert. I've seen desertification happen in Arizona. I was horrified when I drove by some of the rangeland in the early 2000s. Back in the 1970s, it used to be beautiful pasture with a diversity of plant species but now it's an ugly desert full of juniper bushes.

There is no kind of land that turns into a grassland on its own. Grasslands are created and maintained by both grazing animals and fire.

Grazing done wrong can ruin the land but so can no grazing at all. Brittle land you don't graze gets all nasty. Now the land has tufts of grass with dried-up dirt in between each patch. The difference between brittle pasture that is grazed correctly and ungrazed land is striking when you look along a fence line.

One reason people think grazing is bad is that cattle on a ranch with poor grazing management will spread out over a big area and just eat the ice cream and cookies — the grass — and leave the weeds. They are also too spread out to fertilize and evenly eat the pasture.[22]

To get domestic cattle to graze the way wildebeests and bison graze, you need to make them stay bunched. In rough, hilly country, Bud Williams's quiet herding methods work. On the flat plains, portable electric fences are one of the most effective ways to make cattle mimic the behavior of the great herds of bison. The cattle are allowed to graze a piece of ground for a few days and then they are moved.

Today the Nature Conservancy considers grazing part of a good conservation program but we wouldn't have known this without good fieldwork. Pasture rotation systems developed by innovative ranchers can greatly improve the condition of the range.

Being a Pioneer Is Hard

Being a pioneering fieldworker takes dedication. When Allan Savory first presented his observations from his work in Africa, many people thought he was crazy. I remember a long conversation I had with Allan in the 1970s after he had given one of his first talks at a livestock meeting in the United States. He knew he was right, but he had to withstand attacks from people who said he was wrecking ranches. It was a conflict between the academic scientists and a fieldworker. Today many ranchers are using his pasture rotation methods with great success, and some forward-thinking scientists have done studies that show that they work.

Allan Savory calls his land management system *holistic management* because it takes many factors into account. Fieldwork is probably always more likely to be holistic than lab work or mathematical modeling because in the field you can't get away from the whole when a research project starts. A fieldworker might not understand a whole ecosystem any better than bench scientists or model builders, but you're always dealing with it. After a while fieldworkers get a "feel" for the whole system. Bench scientists can study only one or two isolated variables at a time, and modelers can include only the parts of a system they understand well enough to build models for. An example of that is the global warming models, which include fluid dynamics for the ocean and the atmosphere but mostly leave out clouds and dust because scientists don't understand clouds and dust well enough yet to model them mathematically. I'm not saying global warming models are wrong. I'm just saying they can't include all the factors that go into climate. It's true with wildlife, too. No model can include all the strange things that are affecting animals and the ecosystems they live in. Models describe a world that's too pure.

Bench scientists and fieldworkers need to work with each other. Unfortunately, the passion that makes a great fieldworker often clashes with academia. The great fieldworkers are pioneering spirits who go where no one has gone before, and they are impatient with the slow, incremental approach of formal science. Sometimes they make the mistake of attacking academic scientists instead of finding the creative academics who can verify their observations. Some of the best research combines bench science such as DNA analysis with painstaking fieldwork.[23] In the field of botany, the molecular biologists are collaborating with the hands-on scientists who study whole plants. This will provide a better understanding of plant species and help prevent them from going extinct. A similar approach also works well for wildlife.

Need for Hands-on Learning

Why do we have a serious shortage of people going into fieldwork? I think it might go back to childhood, with children staying indoors and playing virtual basketball instead of going outside to shoot hoops. Richard Louv, who wrote the book *Last Child in the Woods,* says kids have "nature-deficit disorder." He talks about one boy who said he preferred to play inside because that's where the electrical outlets are.[24] Today many children have little time for unstructured play outdoors where they can explore and get interested in the natural world. Childhood interests in animals or plants are often the reason a person goes into a career that involves fieldwork. Unstructured outdoor play also teaches valuable problem-solving skills.

I see all kinds of problems with college students who have never had an art class or built anything themselves. This lack of hands-on experience really hurts their understanding of how different things relate to each other in the physical world. My design students, especially the ones who never learned to use a compass or draw by hand, can't make proper drawings. I have been teaching my livestock-handling-facility design class for eighteen years. Since around 2000, the percentage of students having difficulty with the drawings has increased. I think this is due to lack of hands-on experience with drawing in grade school. Last semester I told my students to buy a compass to draw circles with. One girl came up to me after class and said, "Dr. Grandin, I bought a compass and I'm having trouble with my homework." She couldn't figure out how to draw different sizes of circles. When I looked at what she was doing, I found out she had bought a Boy Scout compass and was tracing a circle around its circumference. It's not just students, either. I review drawings from plants around the world and I find the same errors in plans done by draftsmen. Older draftsmen who learned to draw by hand and then switched to the computer do fine. But younger ones who learned to make scale drawings on the computer make basic mistakes like not knowing where the center of a circle is.

The problem is that younger draftsmen have never used a compass to draw a circle and have never built anything with their hands. They do not see their mistakes on a computer-drawn circle because they have never felt where the center of the circle is by sticking the point of a compass in the center. Touch helps the eye to perceive accurately. Oliver Sacks describes a person who was blind and regained vision as an adult. To understand the meaning of things he saw with his eyes, he had to touch the objects he was looking at.[25] I believe that there is something fundamental about the nervous system that prevents the computer mouse from being connected to the brain the same way touch is. Touching and feeling objects are essential for accurate perception.

The field of engineering is becoming overly abstract, too. Colleges and universities are teaching more engineering *science* and less engineering *design*. It used to be that anyone graduating with a degree in engineering knew how to design a combustion engine. Today the majority of engineering students can't do it. An article that ran in *Technology Review* over twenty years ago said that students "may have studied the strengths and properties of various materials or the way gases flow and react in turbines, but they have not necessarily learned how the parts of an engine are designed, manufactured, and assembled — or even how the components work."[26] Fortunately there are some innovative engineering programs where first-year students have the opportunity to design and test a prototype of a new product. Students who participate in these hands-on activities are less likely to drop out of the program.[27]

Making Real Change to Improve Animal Conditions

I think people in general are becoming abstractified. You always hear about autistic children "living in their own little world," but these days it's normal people who are living in their own little world of words and politics. Things have changed. People who wanted to help animals used to study animal behavior. Today they go to law school. That's bad because when everything goes through lawyers you lose sight of the real animals. I am going to give you examples from my experience in the livestock industry, but the principles I have learned would also apply to wildlife issues.

A good example is that the Humane Society in the 1970s used to send representatives to sit in on board meetings of the major livestock associations. That gave the Humane Society direct knowledge of how the livestock industry worked and what things they could change and still stay in business.

In the 1980s, the Humane Society of the United States donated money to fund the development of my center-track restrainer system for meat plants.[28] They would never do that today. Few animal welfare groups would fund something to help reform and improve the livestock industry. As people have become more abstractified they've become more radical, and today the relationship between animal advocacy groups and the livestock industry is totally adversarial.

You see this at every level. Recently I went to a college that has a program on animals and public policy. The only publications they had in their library were animal advocacy magazines. I said, "Look, I think you need to subscribe to *Feedstuffs, Beef, Meat and Poultry,* and *National Hog Farmer.* You need to get the magazines read by the industry." To make policy that will work you need information on every side of the issue.

Dave Fraser, a respected animal welfare scientist at the University of British Columbia, says that to understand an issue you need to read literature that is not from the most extremist people.[29] I believe he is right. Both animal advocacy organizations and livestock groups often respond to complex issues with simplistic and contradictory information. Throughout my career I have observed that on most issues, the best way to solve animal problems is to take an approach that is somewhat in the middle between extremist positions. I tell my students that truth is somewhere in the middle.

Sometimes radical politics are good. I would never have been hired to set up the McDonald's audit system without activists. I use a metalworking analogy to think about animal welfare politics. The big companies are like steel, and activists are like heat. Activists soften the steel, and then I can bend it into pretty grillwork and make reforms.

However, a totally abstractified, legal approach to animal welfare is bad for animals because of the unintended consequences that happen when animal rights organizations try to change things just by passing laws and filing lawsuits. The problem with the legal approach is that it is so abstract that the people have difficulty predicting what will actually happen in the field.

Look at the situation with horse slaughter in the United States. The Humane Society managed to get all the horse slaughter plants shut down in America. Now the old Amish carriage horses and other unfortunate equines are getting transported down to Mexico, where they're worked and starved until they drop dead from lack of nutrition and overwork. If I were a retired Amish carriage horse, would I rather get hitched up to an old pickup truck and get sores and go hungry, or go to a U.S. slaughter plant? I got into a discussion with some of the people trying to shut down the plants once, and I said, "You want to make sure, if you do this, the horses don't have a worse fate." My worst nightmares came true. Thousands of horses have traveled to Mexico, where they were killed by the barbaric process of stabbing them in the back of the neck. Yes, in an ideal world all retired and unridable horses would go to sanctuaries, but we don't live in an ideal world.

We need more animal welfare activists like my friend Henry Spira. Henry was totally rational. He was an old labor negotiator for the National Maritime Union. Henry dressed like a complete slob. He lived in a little rent-controlled apartment with his two cats and a bunch of pieces of cardboard boxes the cats had scratched up. He called them cat sculptures.

Henry was probably slightly Asperger's. He had a little organization called Animal Rights International that was basically his apartment, and he was very effective. People who are interested in wildlife conservation should learn his methods. He got Revlon to stop testing cosmetics on animals. One thing about Henry that made him effective: His word was absolutely good. When Revlon made reforms, other activists kept on bashing them, but Henry said, "You need to back off of Revlon. Maybe they're not perfect, but they're one of the good guys." Henry knew when to apply pressure and when to back off. He made constructive change happen on the ground. Henry died of cancer in 1998. We need more people like him today. Activists need to find out what is actually happening in the field so that true reform will occur instead of the tragic mess of unintended consequences that hurt animals.

Make Wild Animals Economically Valuable to Local People

If you're going to preserve wild animals, you have to make the animals economically valuable to the people in the countries where the animals live. You can't just order people to leave the animals alone if the animals aren't leaving the people alone. If I'm a local tribesman or a farmer and elephants eat all my family's food, I'm not going to feel very charitable toward elephants. Wild animals are going to have to be managed or they're going to die, so incentives have to be in place for local people to want to keep them around. One way to make the animals valuable to the local people is ecotourism. The locals will protect the animals if they can make a living from the tourists.

The Snow Leopard Conservancy of India has developed a program called Homestays where visitors can stay in the houses of local Mongolian herders who have agreed to protect the leopards.[30] The herders make money and the conservancy also pays for sturdy corrals to protect the herders' livestock. Rodney Jackson, the founder of the conservancy, estimates that each project to make a villager's corrals leopard-proof saves five leopards.[31] They also worked with local religious leaders to persuade the herders that wild animals should be protected.

Ultimately there are two ways to manage wildlife. The first is national parks. To have a natural environment, the park needs to have sufficient land for the ecosystem to function. The big predators need more land than the smaller animals. If the land area is too small to support lions and other large predators, other species will overpopulate the area. It is essential to maintain the great national parks such as Yellowstone in the United States and Kruger National Park in Africa. The park systems around the world should be expanded both on land and in the ocean. The second approach is to manage wildlife outside the parks. These are the areas where the greatest losses are occurring. In the less-developed parts of the world, the wildlife will be destroyed by poaching or land development unless the landowners have economic incentives to preserve it. For tourism to work as an economic incentive to protect wildlife, the money earned by the local people must enable them to support themselves and cover costs of damage to their crops and livestock caused by wildlife.[32]

Mike Norton-Griffiths, a conservationist living in Kenya, says that since 1977 Kenya has lost between 60 and 70 percent of its big wild animals in the areas outside the national parks. It was in 1977 that Kenya passed laws making it illegal to hunt wild animals or raise them on ranches to sell for profit. That isn't a coincidence. It was the law that caused the animals to disappear. It made things worse.[33] The large animal advocacy groups are still defending their law.

The law hurt the animals by making their habitat disappear.

Before 1977 wild animals lived in two places: government preserves and privately owned open-range grasslands owned by wildlife ranchers. Once wildlife ranching was made illegal, ranchers couldn't afford to maintain their grasslands. They had to plow up the rangelands and plant crops to support themselves. That law does exactly the opposite of what we need to do to protect the animals. Laws need to be passed that create an incentive for people to take care of the animals. The 1977 law created an economic incentive to destroy the grasslands and deprived the animals of habitat.

African landowners make some of their biggest money selling big-game-hunting safaris. I found lots of websites advertising them. A typical price for a ten-day trip to shoot antelopes and warthogs costs $9,500. This does not include the airfare. A twenty-one-day trip sells for $40,000. These prices will support a lot of local people. I am not a fan of big-game hunting and it is an activity I will not participate in, but sacrificing some warthogs, antelopes, or wildebeests that are held on private land may be necessary to motivate landowners to preserve their land as wildlife habitat.

It took me twenty-five years to learn that money is a major motivator and that economic factors can work either to help animals or to hurt them. Poachers and criminals kill animals because there is a market for products such as ivory, rhino horn, and other animal parts. If activists developed effective educational programs that convinced the younger generation in Asia they should not buy products made from the furs or organs of wild animals, that would benefit animals more than passing laws in Africa. You have to remove the economic incentive to poach wildlife, and you have to create economic incentives to protect wild animals and their habitat.

You can't pass laws against human nature. If you do, the animals will suffer.

Can People Manage Complex Systems?

People involved in wildlife policies need to learn more about how complex systems operate. There's a famous book called *The Logic of Failure,* by Dietrich Dorner, about what happens when people try to manage complex systems. Dr. Dorner is a German psychologist who did a lot of computer simulation studies where he had experts manage complex systems he created.

In one of the simulations people had to manage a West African tribe of seminomadic people called the Moros, who raise cattle and millet. In the beginning of the simulation the Moro are a mess. Their infant mortality is high, their life expectancy is low, they've got famines, and their cattle have tsetse flies.

At the start of the simulation, everyone had a lot of tools to work with: money, fertilizer, the ability to drill more wells, the capacity to set up a health care system, and so on. The people in Dr. Dorner's experiment were very smart and well educated, but almost all of them ended up making the situation much worse than it was to start. After just two simulated decades, Dr. Dorner says, "the Moros were now in a hopeless situation that could only be alleviated by a massive infusion of outside aid."[34]

Not everyone in the experiment failed, however. The people who succeeded at managing a complex system were like field-workers. They were practical, not ideological. They observed what was going on with the Moros, and they constantly asked themselves if what they were doing was working. They especially looked out for surprises — for things they didn't expect to happen. Whenever surprises popped up, they adjusted their actions to take the new information into account.

That is the opposite of abstractification. Dr. Dorner says that "the effectiveness of a measure almost always depends on the context within which the measure is pursued. A measure that produces good effects in one situation may do damage in another, and . . . there are few general rules (rules that remain valid regardless of conditions surrounding them) that we can use to guide our actions. Every situation has to be considered afresh."[35] In my work I call that *animals will throw you a curve ball.* Animals have a mind of their own and they will sometimes behave in ways that are difficult to predict. You have to pay attention and change your design if you need to.

Teaching Students How to Observe

I have lived in Fort Collins, Colorado, for eighteen years and I have noticed how the behavior of the Canada geese has changed. They used to always stay together in large flocks, but today I see more and more single pairs around town. A pair may be grazing on the bank's front lawn or nesting under the steps of our campus building. Since they are not hunted in town, they are spreading out and flocking less. People who are good at fieldwork immediately notice these important details of how animal behavior changes as the conditions change.

I have an exercise I do in my livestock-handling class to try to teach students how to understand animals through observation instead of abstractification. I have them capture and handle imaginary animals. There's a fire-breathing dragon in the foothills, I tell them, and your job is to bring him back alive without having him burn up Fort Collins and our university. Or there's a ten-foot-high daddy longlegs out in a field. Your job is to bring him back to the lab alive without breaking his legs. I tell them they can't use anything made-up. They can't get a magic ray gun and shoot the dragon and bring him back. They have to work with things we have now in the real world.

The first reaction of the veterinarians in the class is always, "Let's shoot him up with drugs."

I always say to them, "You don't know anything about this dragon and you're going to shoot him full of ketamine? You don't know if tranquilizing drugs will work, or if they'll kill him, or if they'll just make him mad. You might get him enraged and he could trash Fort Collins."

Then I tell my students, "Why don't you just watch him a little while?"

One thing I might want to know if I were trying to catch a dragon is how long it takes to regenerate his fire. I might get in a helicopter and tweak him just a little, not to hurt him but to get him to blow out his fire. If it takes a couple of hours for the dragon to regenerate his fire, that's the safe time to catch him.

With the ten-foot-tall daddy longlegs, students will instantly say, "Let's go out with horses and rope him."

I say, "You'll break his legs."

After that, they're usually stumped. I remind them, "Remember, insects shut down in the cold because they're cold-blooded." So I tell them we should wait until the temperature drops to catch him, or, if we have to catch him now, we will have to chill him. Then I get the students brainstorming on how to chill him without freezing him and causing death. They think of all kinds of tents that could be put around him and hooked up to an air-conditioning unit for chilling.

After that I ask them how we're going to transport him to Denver without breaking his legs.

Usually they have no idea, so I give hints. There's something very common that will work, I say. We have lots of them here in Colorado.

The answer is refrigerated trucks. Once I've got him chilled I can fold him up, put him inside the truck, set the refrigerator unit to about 45 degrees, and transport him to Denver.

That's the way a fieldworker thinks.

9 Zoos

WHEN I WAS A CHILD going to the zoo, the monkeys and the big cats and all the other animals lived in enclosures that looked like tiled bathrooms. There was nothing for them to do. I remember one poor elephant that just stood in one place and swayed back and forth. It was terrible.

Things weren't any better in the 1970s when I lived in Arizona. For my animal behavior class, we had to go to the zoo to observe an animal, but there was almost no animal behavior at the zoo to observe. There was no point observing the coyote because all he did was endless circling stereotypies. The big cats slept all day because Phoenix was so hot. I finally chose the oryx antelopes because they were one of the few animals that lived in a large enclosure where a group of animals had opportunities for somewhat varied behavior. I watched two males in adjacent pens sparring with each other by sticking their horns through the mesh of a chain-link fence. That was the most interesting thing any animal had to do with itself in the entire zoo.

The first feeble attempt that zoos made to improve the cages was to add some ropes for the monkeys to swing on. So the monkeys lived in tiled bathrooms with ropes while all the other animals went on sleeping all day or pacing in circles.

Fortunately, many zoos have made a lot of progress creating good environments for their animals since then. One meta-analysis of twenty studies of enrichment programs found that the programs have reduced stereotyping by 50 percent on average.[1] That is a very large effect.

Zoos have also developed a good set of commonsense criteria to judge an animal's welfare:

- Is the animal acting normal? (Does it have abnormal behaviors such as repetitive stereotypies, abnormal aggression, or self-injury?)
- Is the animal busy doing different things? (Does it have a "wide repertoire of behavior"?)
- Is the animal confident? (Does it freely move around its enclosure without acting afraid?)
- Does the animal act relaxed when it's resting? (Or does it act hyper-vigilant and on guard?)

These four questions are a simple, commonsense "audit" of captive animal welfare, and zoo employees who know their animals well can easily ask and answer them. Zookeepers who understand the purpose of the questions have been able to collect a lot of useful information about what captive animals need to have a good mental and emotional life.

What needs to happen next is that all zookeepers should learn to identify abnormal behavior. I visited an aquarium where a single dolphin was swimming around and around his pool in a repetitive pattern that never varied. He swam the entire length of the tank along the bottom, and then swam up at a 45-degree angle to the corner across the pool. When he got to the top corner, he turned and swam the length of the tank at the surface, and then he followed a second diagonal at a 45-degree angle down to the corner at the bottom of the tank, which put him back where he started. A lot of animals develop circular stereotypies where they move in one plane, but this dolphin had developed a figure eight that used all three dimensions of the tank. His path was unusual enough that the keepers were not aware that this was an abnormal stereotypy. The best zoos train their staffs, but there are still too many zoo managers who do not fully recognize the importance of providing enrichment to prevent abnormal behavior, partly because they don't recognize abnormal behavior when they see it.

Other zoos have gotten sidetracked by bad ideas about animal welfare. One of the most common is the "back to nature" approach where the goal is to make the enclosures as close to the animal's natural habitat as possible. That sounds logical until you stop to think that "nature" in a zoo is nothing like nature in the real world. Real nature means predators or prey, disease, hunger, and danger. "Zoo nature" doesn't have any of those things except disease, and a sick zoo animal gets immediate treatment from a veterinarian. The result is that some zoos have spent a lot of money building fancy enclosures that appear natural to people, but are just as boring and painful for the animals as a barren concrete cage. I remember one tiger exhibit that looked really pretty with lots of rocks molded from concrete. There was absolutely nothing for the tiger to do. The enclosure was visually stimulating for people, but it was a barren environment for the tiger.

Prey Species Animals in Zoos and FEAR

I'm going to start by talking about prey species animals and predator animals separately because different animals have different weightings for the blue-ribbon emotions, and the main difference is between predators and prey species animals. For prey animals the most important emotion system zookeepers need to think about is usually going to be FEAR; for predator animals the most important system is likely to be SEEKING.

Overall, I have observed that the herbivores such as deer, nyala antelopes, Cape buffalo, and other hoofed stock adapt much better to life inside a zoo than the large predators do. These species spend huge amounts of time in the wild eating grass and other forage, and it is easy to provide lots of forage at a zoo, which goes a long way toward satisfying their SEEKING system.

All prey species animals are high-fear. So with prey animals, your first job is to make sure you do not create an enclosure that continuously activates the FEAR system in the brain. I have found that many zookeepers don't realize that the physical environment is just as important to FEAR as it is to SEEKING. Zookeepers make the same mistake chicken farmers make. They assume that as long as there aren't any predators around, prey animals feel safe. But that's not the case. A hen needs a secluded place to lay her eggs regardless of whether she's ever even seen a fox. In zoos, many small prey species animals need objects they can hide underneath. It's not enough to have a secluded corner they can retreat to or a barrier they can go behind. These animals have evolved to hide from aerial predators, and the only way to hide from an aerial predator is to get underneath a solid object.

Primates have the opposite need. They feel exposed and scared if they do not have high places to sleep at night. In the jungle, a tree is a safer place to sleep than the ground, and many primates are hard-wired to seek safety in treetops.

To understand what an animal needs to feel safe, its behavior must be observed in the wild. Where does it sleep? Where does it go when it is threatened by a predator? What are the features of places an animal chooses? To keep the FEAR system from being chronically activated, these features have to be built into the enclosure.

Old Thing in a New Place

It's not easy for normal human beings to understand animal FEAR — especially the fear felt by prey species animals in zoos.

A couple of months after *Animals in Translation* came out, I got a call from a zoo. They were having a horrible time getting their antelopes shifted from the barn where they slept at night to the exhibit area each morning. To get to the exhibit the antelopes had to walk through a fenced-in alley, and on some days — not every day, just some days — the antelopes were balking. They would not walk through that alley. Something was scaring them.

The reason the zoo called me was that they'd read *Animals in Translation* and they'd done everything the book said to do, but their animals were still going crazy. They couldn't figure it out.

So I went to the zoo and talked to the keepers. They said they thought the problem had to be something to do with novelty, but there wasn't anything new in the alley, and they couldn't see anything different about the days the antelopes walked calmly through the alley versus the days they balked. They'd been over it and over it. Everything in the alley had always been in the alley and was still in the alley on the balky days.

I went in and took a look. The keepers were right. There wasn't anything new inside the alleyway.

The problem turned out to be that there was something old in a new position. The alley was enclosed by a chain-link fence with an electric box mounted on it that was supposed to have a warning sign attached to its side. However, the employees didn't have the screws to put it up, so they'd just leaned the sign up against the fence down below the box.

The sign wasn't very stable propped up like that, and on some days it got knocked over. That was the problem. The front of the sign was black and white, and the back of the sign was bright yellow. When the sign tipped over, the yellow back was showing. Yellow really attracts an animal's attention. That's because the only animals that have full color vision are primates and birds. All the rest have dichromatic vision, which means that they see two main colors — bluish purple and yellowish green — and they don't see red. Anything yellow in the environment will "pop"; they'll notice it right away. Every time the sign got knocked over, the yellow back was visible on the ground and the antelopes reacted.

The fact that the sign was out of place was probably even scarier. Most days the sign was in its proper spot, leaning up against the fence. That's where the antelopes expected it to be. Then all of a sudden it would be lying on the ground and it would be yellow, too. That's a big change for an antelope.

The reason the humans didn't pick up on it was that to humans, a sign is a sign. It's so hard for normal people to think like animals. A good analogy would be if you started to walk into your living room and you saw that the sofa had been turned upside down and spray-painted a different color. That would definitely scare you.

Within a species, different individuals have different hyper-specific fears, too. Most fears are learned, so different zoo animals with different histories can acquire different fears over the years. Those fears are going to be hyper-specific because they are stored in the animal's memory as pictures or sounds. Usually the animal is afraid of something neutral that it was looking at or hearing when the scary event occurred, which makes it even harder for the human keepers to analyze. I saw a case of an elephant who feared the sound of diesel engines because he had been shoved around by a big tractor. Car engines did not bother him. Fortunately, in that case the keepers knew what the problem was and knew to keep diesel-powered equipment away from him.

The only way to deal with hyper-specific fears is to figure out what *exactly* is scaring the animal through close observation. Once you identify the frightening person or thing, you have two choices: try to habituate the animal to the object or remove it from the animal's environment. If the object is something easy to eliminate, then eliminate it. But if it is a common object like a white shirt, habituation is recommended because we cannot rid the world of white shirts. Getting an animal over its fears is easier in the less flighty species.

Training Animals to Cooperate with Veterinary Procedures

Another thing captive prey species animals need is trained staff to habituate them to veterinary procedures and other necessary forms of handling. All captive animals need this, but prey species animals absolutely must have keepers who know how to habituate and train high-fear animals.

The idea that you can train high-fear flight animals to cooperate with veterinary procedures was new when I first started working on it. About ten years ago the nutritionist for the Denver Zoo, Nancy Irlbeck, wanted to do a study to determine how much vitamin E the zoo's four nyalas had in their blood as part of a larger study of their nutritional health. (Nyalas are a small South African antelope.) The trouble was that stress suppresses vitamin E levels, so if you stress the animal it's impossible to get an accurate reading.

Nancy called me and asked, "How do we get a no-stress blood sample out of an antelope?" I told her there's only one way: You have to train the animal to voluntarily cooperate during the blood test. There's no other way to do it.

People thought that was crazy. Nyalas are super-shy, and people who knew what they were like were saying, No, no, no, there's no way to train these animals, because they will go berserk and kill themselves while you're trying to manipulate their behavior.

Oh yes, there is. I worked with some great students, and we trained those animals to stand still calmly while a lab technician drew blood.

The secret to working with a high-fear prey animal like a nyala antelope is to start with a very long habituation phase. Before I could even think about training the nyalas to go inside a wood box and hold still for a blood test, we had to spend ten days just getting them used to the sound of the wood door sliding up and down. The first day we could move the door only one inch because if we'd moved it any more, that animal would have gone splat on the wall. That's what people were worried about: that we'd startle the animal, and it would kill itself in its panic.

The other danger we had to worry about was the fact that hyper-flighty animals can form fear memories so strong that they never get over them. If we had let them panic just once at the start, that could have made it impossible to train them.

We quickly discovered that the key to successful habituation of a high fear prey animal is to *never* push the animal past the "orienting" stage. Orienting is what animals or people do when they hear a new sound or see something strange. They stop what they're doing and look toward the new thing. During the orienting stage, a high-fear animal makes a split-second decision: "Do I panic and run, or do I keep looking?" The choice is either to SEEK or to allow FEAR or RAGE to take over.

I think it's possible that prey species animals have a much faster path between orienting and FEAR than non-prey species animals. One time when I was driving on the freeway a big two-by-six board slid off a trailer just ahead of me in the next lane over and came flying diagonally across the road in my direction. My attention locked onto the board like radar, and it was as if everything slowed down. The board looked like it was floating on the road toward my car. My driving became super-controlled and I moved over to the breakdown lane where the board slid on the ground so I could straddle it between my front wheels. During this time I was in a state of pure concentration with no emotion. As soon as I realized I was safe, my FEAR system switched on and I felt terrified, but not until that moment. If I had panicked the instant I saw the board coming at me, I would have had a horrible accident.

I've never seen a prey species animal do anything like that. High-fear prey animals go straight to terror if you push them past the point of orienting to novel stimuli, and they can hurt or even kill themselves in panic. So we stopped the habituation program the instant an antelope oriented to what we were doing. When we moved the door one inch on the first day and the antelopes oriented to the sound, we stopped the program for that day. On the second and third days, we opened the door a few more inches. The animals oriented, and we stopped for the day. It took ten days altogether to habituate the antelopes to the door being opened. By then we could yank the door open quickly without drawing the animals' attention. They didn't orient no matter how far or how fast we opened the door.

At that point we could begin a combination of training and habituation. We used special yummy treats to lure the antelopes inside the box, and we stopped work whenever they oriented to some aspect of the box or the training.

We spent fifteen minutes a day for weeks training them to go in the box, keep their legs still, and let us get a blood sample out of their legs. We used no sedation or drugs of any kind.[2] People were amazed. The only way anyone had ever been able to get blood work done on the big prey animals was to shoot them with a tranquilizer dart gun and collect the specimens while the animals were unconscious. Zoos were darting their animals at least once a year, and it was horribly traumatic. I've talked to one zoo vet after another who has done lots of darts. Many end up quitting because the animals hate them so much.

Even after we finished training the antelopes, the veterinarian who had darted them in the past was never able to handle them. A strange vet could examine them, but not the dart gun man. They recognized his voice, his appearance, and his gait. These animals aren't just afraid, either. They're angry. They hate the mean, nasty vet who does all the horrible procedures.

Zoos don't dart all their animals, just the big ones. To do procedures on the smaller animals they physically restrain them, which is just as terrifying for the animal as being darted. With a small animal, like an otter, a keeper catches the animal with a fishnet and then holds it down until the vet is finished. To restrain a medium-sized animal like an antelope, the keepers take a big sheet of plywood and crowd the animal up against a wall in the barn and hold it there. Using restraint triggers RAGE, and the veterinary procedure triggers FEAR, so the smaller animals hate the vet as much as the big animals do.

We still had to be very careful after the antelopes were fully trained. The nyalas were mostly female, and they were Miss Hyper-Specific. If you trained them to tolerate one thing and then you deviated from that one thing, they panicked. One day there were some men fixing the roof of the barn where the nyalas slept at night, and the antelopes freaked out and slammed into the chain-link fence. Fortunately they were OK. They panicked because they had learned that people in front of their exhibit were safe and also that people inside the barn were safe so long as the dart gun vet wasn't there. But people on the roof were something totally new and scary.

The zoo was happy with our results, so they decided to train their bongo antelopes next. I trained the student trainers. While I was working with the students I had an idea. There was a keeper at the zoo, Megan Phillips, who was in charge of collecting blood from the antelopes once a month whether they needed it or not just to keep them trained to the procedure. One day I said, "Megan, what are you doing with that blood?"

She said, "We just put it in the freezer."

So I told her, "Send some of that blood to the lab and get a glucose, CPK, and cortisol done and send the bill to me." Glucose, CPK, and cortisol are all related to stress, and stress hormones weren't part of the study. I decided to tack on a cortisol test because the blood was already there, stacked up in the freezer. The levels came back incredibly low, almost at the level of cattle that are asleep, even though each animal had stood in the blood-testing box for twenty minutes. The scientific literature had values from netted or darted animals that were three and four times higher than what I got in our trained animals. Researchers were calling those elevated values "normal" because everyone always got those values when they drew blood from captive antelopes. But the reason everyone always got those values was that everyone was drawing blood from terrified animals.

That's one of the big problems in zoo and wildlife veterinary medicine. A lot of people who work with wild species don't understand that if you throw a fishnet over an animal and hold it down so you can get blood drawn, you stress the crap out of it. I've had people tell me, "It can't be that stressful; we just hold them down for thirty seconds." I say, "Yeah, a mugging on the subway only takes thirty seconds, and it's real stressful." Training the animals to cooperate with veterinary procedures is much more humane.

We wrote up our results in a paper we called "Low-Stress Handling of Bongo" and submitted it to a journal. One of the reviewers objected to the title. He said "Low-Stress Handling of Bongo" was judgmental because it implied that the regular methods of drawing blood were stressful for the animals—which they were. We had to change the title to get the article published. We finally called it "Crate Conditioning of Bongo (*Tragelaphus eurycerus*) for Veterinary and Husbandry Procedures at the Denver Zoological Gardens."[3]

Hyper-specific Pronghorn Antelopes

A few years after I trained the nyalas I was hired to help train pronghorn antelopes at another facility. These pronghorn fawns had been hand-raised, but they were still very flighty animals. That's their nature. When I got there they were in outdoor pens made out of plywood, and they would panic when Canada geese flew over the pens in a different direction from the way they usually flew over. Pronghorns were even more hyper-specific than the bongos or nyalas.

We eventually got them so well trained that a student could walk right up to an antelope and stick an IV into its jugular without the slightest resistance from the animal. We didn't need a blood-testing box. Then one of the pronghorns got sick and needed an antibiotic shot in the shoulder, not the neck. When that same student tried to give the shot, the antelope went berserk. It had been trained for "IV in neck," not for "shot in shoulder," and it hadn't developed a general category called "getting a shot." The antelopes' hyper-specific brains detect and react to small differences.

On the other hand, the pronghorns had developed some general categories for small neutral objects. I noticed that they had no reaction to novel coffee cups or drink bottles trainers carried into the enclosure. They had seen so many different drink containers that they had probably opened up a new file folder in their brains labeled "small things in people's hands are safe."

They were much more fearful of large novel objects compared to small novel objects. We had to introduce a new piece of plywood or a large chest very carefully to avoid a crackup on the fence. One of the reasons large objects are scarier is that the predators that eat the pronghorns are large. The pronghorns are naturally more inclined to acquire a fear of big things.

We didn't use any negative reinforcement or punishment at all when we trained the antelopes. We couldn't. It would have wrecked them the way sacking out a horse wrecks an Arab horse. Negative reinforcement and punishment have always been used by some trainers with big, not-so-fearful animals like elephants that can take a lot of abuse without panicking or getting sick the way a dolphin would. It's still a bad idea with those animals, because elephants never forget, and elephants that have been abused in the past may attack a trainer. But you can get away with it. Using either negative reinforcement or punishment with a high-fear animal will destroy any chance you have of training that animal to cooperate.

Captive Predators and the SEEKING System

Captive predators have lower FEAR but higher SEEKING needs than prey species animals. One major hurdle to providing opportunities to SEEK is that in the wild a predator kills and eats live prey. Western zoos won't give a predator a live animal to eat. In England, it's even been made illegal to give live fish to otters and sea lions.[4]

In one developing country zookeepers did give the tigers live cattle to kill and eat. This was probably an enjoyable experience for the tigers, but it is a practice that I am totally against because the cattle would really suffer. The poor cattle would have no way of escaping from the small enclosure and their FEAR and RAGE systems would be fully activated. All mammals and birds would suffer if they were used as live prey to feed zoo animals. Research is needed to determine if fish suffer the same way mammals and birds do. My opinion about live prey is that feeding live worms or insects is ethical because insects do not have the ability to feel pain. If research shows that fish have sufficient nervous system capacity to suffer, I would recommend euthanizing them before they are fed to the animals.

Animals are adaptable. What zoos need to do is find other methods of satisfying a captive predator's SEEKING needs. Of course, that's not easy with the large predators. It is much more difficult to build exhibits for large animals that provide a natural enriched environment than it is for small animals.

Dr. Hal Markowitz was the first ethologist to create alternative sources of SEEKING for captive animals, including the big predators.[5] He installed an acoustic-prey device for a sixteen-year-old African leopard named Sabrina at the San Francisco Zoo. When Sabrina climbed up to the top of the left-hand wall of her enclosure, she triggered a motion detector that turned on a speaker that played a recording of bird song. The first speaker triggered a second speaker, which was a few feet away and farther down and also played bird song. The second speaker triggered a third speaker another few feet away, and finally the third speaker triggered a fourth speaker on the bottom of the cage and on the opposite wall from the first motion detector. When Sabrina jumped down to the fourth speaker, her motion triggered another motion detector that released food into a food chute.

Sabrina invented a few other ways of chasing the bird song, too. Normally she chased the four sequential sounds on a diagonal from the top of one wall of the cage to the bottom of the other wall. If she was feeling especially active, though, she didn't bother with the chase. She just shook the top branch so hard that the movement triggered the second motion detector clear over on the other side of the cage and the food dropped out of the chute. She also had several shortcuts she could use to go straight from the first speaker to the food without stopping off at each of the speakers.

Dr. Markowitz was also the first ethologist to take a scientific approach to enrichment. He took baseline measures of Sabrina's behavior before and after the zoo put in the acoustic-prey device and found that she was much more active and cheerful-seeming once she was able to chase acoustic prey.

Gus the Polar Bear

People have always noticed that the big predator animals develop some of the worst stereotypies in captivity, but no one knew why. There were different schools of thought. The most common hypothesis was that stereotyping in the big predators was a stunted form of hunting. Many researchers also wondered whether the big animals end up pacing or swimming figure eights because they need more exercise than they can get inside a zoo enclosure. Both of those explanations make sense because the big predators cover a lot of territory when they track prey (exercise), and, in the zoo, they usually stereotype the most intensely right before they get fed (need to hunt). Dr. Georgia Mason at the University of Guelph found a stereotypy peak before feeding in more than 70 percent of twenty-one predators she surveyed, and there is also evidence that the big cats stereotype when they see "potential prey," including ponies and children running past the cage.[6]

In 2003, Dr. Mason and her colleague, Ros Clubb, published a major study of stereotyping predators in *Nature*, which found that neither hunting nor activity level predicted stereotyping in captivity. What predicted stereotyping in captivity was *ranging*, which is how far an animal travels in a normal day.[7] Ranging isn't the same thing as hunting, because a wide-ranging animal roams or "travels" even when it's not looking for food. The best example is polar bears. Polar bears are one of the farthest-ranging animals we know of. They travel five and a half miles a day and are fantastic swimmers that can swim for hours at a time. The longest polar bear swim a scientist has recorded is two hundred miles.

They also stereotype horribly inside zoos. Dr. Mason and Dr. Clubb found that this was typical. The wider-ranging the animal, the more the animal stereotyped in captivity. For example, lions and leopards both hunt the same amount in the wild, but lions in zoos do five times the amount of pacing leopards do. That's very unexpected, because leopards living outside the zoo are much more active than lions. Lions are big slugs. They're active only a little more than two hours out of every twenty-four, whereas leopards are active twelve hours a day.

But inside a zoo, it's the lions that pace like crazy, not the leopards. The difference between lions and leopards is that lions range much farther than leopards. It's the nomad animals — the animals that don't really have a home or even a home territory — that develop the most intense stereotypies inside a zoo. These animals are the opposite of homebodies. A homebody animal, like a red fox, can have a home territory that is less than a square kilometer. That's so small the fox can cross the whole area in just a few minutes. Foxes do well inside zoo enclosures.

Nomad animals like African wild dogs and wolves don't do well, because they are roamers. African wild dogs don't spend more than two nights in the same place, and wolves don't spend more than a few nights in one place before moving on. They develop a lot of stereotypies in captivity.

A zoo enclosure can work pretty well for a fox, but it's hard on a polar bear or a wolf. No matter how nice and roomy a zoo is, it's still a home. Nomad animals don't want a home.

Whether or not we should even have nomad animals in zoos is a big issue. A lot of people think we may not be able to keep these animals inside zoos. Captive polar bears have been recorded spending anywhere from 16 to 77 percent of their recorded hours stereotyping,[8] and I think that if you can't solve the stereotypy problem, they shouldn't be there. But you also can't just send an animal that has spent its whole life in captivity and doesn't have any survival skills out to the wild to die. So with most stereotyping captive animals, you're going to have to figure out some way to provide mental stimulation inside the environment they're living in now.

Since zoos don't have enough space to have big predators ranging loose five and a half miles a day, what can they do to create good mental welfare for animals like polar bears?

One approach I've seen work with the famous polar bear at the Central Park Zoo used the brain's PLAY system. Gus the polar bear had a figure-eight swimming stereotypy he was doing for 80 percent of his waking hours. In the middle 1990s *Newsday* ran an article about Gus, and a lot of people got very upset about him.

In 1994 the zoo hired a behaviorist to help them figure out how to make Gus and his companion in the exhibit, Ida, happier. They tried a lot of different things, a few of which made Gus stereotype even more. But by 2005 both bears were spending only 10 percent of their day doing stereotypies, which is a huge reduction.

I saw Gus in 2005. The zoo had put a bunch of barrels with different levels of buoyancy in his pool, and Gus was having a blast jumping on the barrels and pushing them underwater. The barrels would pop back up out of the water and he'd jump on them again. He looked like a kid at a swimming pool with a diving board, jumping off the board into the water, surfacing, climbing out, and jumping off the diving board again.

I watched him play with the barrels for probably forty-five minutes. That sounds like a long time, but his play definitely wasn't stereotypic. Stereotypic behavior is always exactly the same, and the barrels made it impossible for Gus to do the same thing over and over again, because they popped up in a different position each time he jumped on them.

I don't know whether the barrels also activated Gus's SEEKING system, but since animals don't play when they're depressed, angry, or afraid, his mental welfare while he was playing with the barrels was decent. He was definitely better off playing with water barrels than swimming figure eights all day long. I recommend that zookeepers give their animals lots of opportunities to SEEK and to PLAY — especially the nomad animals. Allowing nomad animals to continually pace is not acceptable. Nomad animals should be kept in zoos only if their environment is sufficiently enriched with play, animal companions, or interaction with keepers to prevent pacing and other stereotypies.

Food for the Brain — Things That Work with All Animals

Predators may have higher SEEKING needs than prey species animals, but all captive animals need enriched enclosures that give them lots of SEEKING activities.

One of the best ways to stimulate any animal's SEEKING system is to have them "work" for their food, which was what Dr. Markowitz did with Sabrina the leopard. Dr. Markowitz created food enrichments for primates, too. In 1972 the Oregon Zoo in Portland hired him to create a better environment for their gibbons. The monkeys were living in an old-fashioned concrete cage with plain walls and bars up front. Dr. Markowitz installed a food delivery device high up in the cage that let the gibbons work for their dinner. The device had a light at the top of the cage, a lever on one side, and some small platforms built across the cage like steppingstones in midair. Whenever the light went on, the gibbons could pull the lever and food would be released on the opposite side of the cage. Then the gibbons would hopscotch across the stations to get to the food.

The gibbons didn't have to work for their food if they didn't want to. Their keepers gave them any leftover food they didn't get by pulling the lever at the regular feeding time at the end of the day. But they liked working for food, and they spent the next seven years jumping around the top of the cage to "forage" for their food, instead of eating it on the floor at mealtime when the rest of the primates got fed.

Dr. Markowitz's system worked out great because zoo visitors seemed to like watching the gibbons pull the lever and jump across the cage as much as the gibbons liked doing it. The zoo connected a coin slot to the contraption so visitors could deposit a dime that turned on the light and started a new "trial." In one year the zoo collected $3,000 — 30,000 dimes — from visitors who wanted to turn the light on and watch the gibbons pull the lever and leap across the cage to get their food.

Dr. Markowitz's next invention was a token system for the Diana monkeys at the zoo. It had the same leaping stations built in across the top of the cage, but in the new system, when the light came on the monkeys pushed a lever to get some plastic chips. They could use the chips to "buy" food from an automat that Dr. Markowitz put into their cage.

The Diana monkeys developed a lot of new and varied behaviors besides pulling a lever and jumping across the stations to collect their tokens. Some monkeys spent their tokens on food as soon as they got them, others gave their tokens away, and some monkeys stole tokens from their cage mates. Dr. Markowitz tells a funny story about Butch, a teenage Diana monkey whose mother started pushing him away and taking the food he'd just paid for at the automat for herself. She did it so often that Butch came up with a way to trick his mother by pretending to drop a token into the automat but palming it instead. His mother would come over to steal his food and find the automat empty. When his mom gave up and left, Butch would put the token in for real and get his food.

Ethologists have observed animals tricking each other in the wild in order to get food. They call this kind of behavior *tactical deception*. Dr. Markowitz's behavioral enrichment devices are very good examples of the fact that a totally artificial situation can produce normal behaviors in a captive animal. No Diana monkey living in the wild has ever tricked his mother out of stealing food from an automat. Dr. Markowitz writes, "Where captive environments cannot include the replication of natural contingencies, unnatural ones may serve to provide animals with power."[9]

Many other researchers have found the same thing Dr. Markowitz did: animals will choose to work for food instead of just getting the food handed to them. The easiest, most surefire way to increase the SEEKING emotions and behaviors is to stop feeding zoo animals in food bowls or feed troughs as if they were pets or farm animals. They're not pets or farm animals; they're wild animals, and they're built to go out and find food. Many, many studies have found that captive animals will choose to work for food instead of just having it handed to them. Wild animals don't want a free lunch.

The reason they like working for their food is that it feels good. That's because in all of the studies "working" actually means SEEKING. The animal has to forage for hidden food in its enclosure, or manipulate a puzzle feeder (a kind of container with food inside), or chase acoustic prey. All of these activities activate the SEEKING system. They let the animal hunt.

Another great thing about using food as enrichment is that animals never get bored with food. Animals habituate very quickly to everything else, but they stay interested in food and will work for it even when they aren't hungry.

The SEEKING system protects both animals and people from stressful negative emotions. Dr. Susan Mineka, a psychologist at Northwestern University, did a study showing that baby rhesus monkeys that worked for their food were less fearful than the babies that didn't. The SEEKING babies were also less upset when they were separated from their cage mates.[10] Of course, you have to make sure to provide enough enrichment devices to go around. If animals get into fights over enrichment devices, you've introduced a new source of stress into their lives, not enrichment.

Dr. Markowitz's contraptions worked well for the animals, but they weren't very practical because they were always breaking down. Part of the reason his things were so complicated, though, was that he was trying to enrich a totally barren zoo environment on a very low budget. There wasn't enough money to build new exhibits, so he turned the whole cage into a kind of

Skinner box. It was the 1970s, when everything was pure behaviorism, and people called his approach *behavior enrichment.*

Nowadays most zoos have been able to replace their old concrete prison cells with soft environments featuring real trees, water, interesting substrates to explore, places to hide for prey species animals, and so on. In a well-designed soft environment you don't need a fancy acoustic-prey setup. You can rig up a simple pulley and cord and have the keeper stand outside the enclosure dangling giant cat toys — like a big cardboard box filled with meat — for the tigers to play with. The big cats will chase it, bat it, and pounce on it. Their behavior is exactly like that of a housecat chasing one of those feathered wands you can buy at a pet shop. To keep giant tiger toys really interesting, you can use two pulley systems so the keeper can move the box sideways and up and down. I even visited one zoo where they had children make piñatas the lions and tigers could "kill."

Both the cats and the kids had a good time.

Using Positive Reinforcement to Turn On SEEKING

Another excellent way to turn on SEEKING in zoo animals is to use positive reinforcement to train them to cooperate with all zoo routines, not just veterinary procedures. As I've talked about in the other chapters, positive reinforcement turns on the SEEKING system by teaching the animals to anticipate a reward when they hear the click or when they perform a behavior on cue. They are SEEKING rewards by doing what they've been taught to do.

The side benefit is that positive reinforcement training is much safer for the humans, too. Elephant training used to be a dangerous job that involved a lot of negative reinforcement. The trainer would try to push the elephant into the position he wanted it to be in, and the elephant would either get into the position in order to get away from the pushing (negative reinforcement) or get mad. If the elephant got mad, the trainer could be injured or killed.

Now zoos are using positive reinforcement and clicker training to train their elephants. The trainer charges up the clicker and shapes the elephant's behavior a little bit at a time. If you want to train an elephant to back up to a fence and pick up her hind leg and place it on a stand so the keepers can file her nails, first you click her for turning her head away from you, then you click her for turning her head and shoulder away, and so on. You don't have to push and shove and yell to try to get the elephant into the right position. It's fun for the elephant and much safer for the human.

Zoos are also using positive reinforcement to do target training, which is an extremely useful behavior for any animal to learn. In target training, the animal is trained to touch his nose to a ball on the end of a stick. That's the target. Once the animal has learned to touch the end of the stick he will tend to follow the ball, and keepers can lead the animal where they'd like him to go instead of trying to drive or shoo him from one place to another. With more extensive training, a zoo can teach its animals to move back and forth between the exhibit enclosure where they spend their days and the barn behind the exhibits where the animals eat and sleep at night. On cue, each group of wild animals moves smoothly from one place to the other. Before, some animals refused to shift and keepers resorted to poking, shouting, or deeply invading their flight zone to force them to move.

A Treat Must Really Be a Treat

To motivate a wild animal to perform for a reward, the reward has to be super-desirable. I have seen firsthand that an animal will not perform for a treat that is not really a treat. At one zoo, I visited with a gorgeous, silky smooth okapi that loved to be stroked. The zoo nutritionist had forbidden the use of what he considered unhealthy treats. The only treat he allowed this animal to have was bok choy, which is like giant celery. The okapi hated it, and the training program didn't work.

When I first started working with the antelopes, the keepers suggested that we should use their regular food as a reward. I said you need to use a real treat that will be like cake and ice cream to the animal. To figure out what the nyala and bongo antelopes really craved, we laid out samples of every type of produce in the zoo's kitchen in a row and let them pick. The nyalas picked yams and the bongos really liked spinach. So that's what we used for the reinforcers in our training program.

Zoo nutritionists are often concerned that the treat will be bad for the animal's health, especially if it is something unhealthy like grain or marshmallows. To get around this problem, you can use very small amounts. Animals can be rewarded with as little as a teaspoon of grain. One keeper I met learned that an elephant will perform for a single miniature marshmallow.

Natural versus Wild Behavior

Not everyone believes in training programs for captive animals. The people most likely to object are either "macho" keeper types who enjoy forcefully wrestling with an animal, or the more theoretical zoo management and wildlife purists. Wildlife purists don't like zoo training programs because they believe that training interferes with behavioral conservation, which means conservation of a wild species's behavior, not just its physical features. They say training programs take the wildness out of wild animals and change their basic natures.

However, captive animals have already had their behavior changed before you do any training. Most animals that have been raised in captivity cannot be returned to the wild. They don't have the skills.

Also, captive animals have to have medical care. Even if a zoo decided to go all-natural by withholding veterinary care, they'd still have to follow state and federal laws requiring that certain disease tests get done anytime they moved an animal to another zoo. If zoos don't train their animals to cooperate calmly with veterinary procedures, they have no other choice but to stress them horribly by forcing them to undergo the procedures against their will. Training is much more humane.

Although a training program reduces the animal's natural fearful behavior in one setting (veterinary procedures), training programs can actually help preserve the wild-type flighty genetics. My friend went to a wildlife research station, which is a kind of research zoo without exhibits that had untrained Rocky Mountain bighorn sheep. When the sheep had to be blood-tested, the researchers threw nets over them and sat on them to hold them down. My friend helped sit on the animals and told me that the hearts of some of the sheep were pounding so hard that they felt like they were jumping out of their chests.

Just a few weeks later, a high percentage of those sheep were dead from disease. This had been a healthy flock with no disease problems until they had freaked out while being restrained. FEAR lowered their immune function. This is a horrible way to handle beautiful bighorn sheep. Very likely the sheep with the most flighty high-fear genetics were the ones that died because they were the ones most severely stressed by the experience.

A pronghorn antelope that will let you take a blood sample out of her jugular vein but hit the wall if you try to give her a shot of antibiotics in her shoulder isn't going to lose all her wild behavior because the zoo trained her to cooperate with the veterinarian. Animals are specific in their thinking. They have one file folder inside their brains for "we go do vet stuff and get fed treats" and another file folder for "we're doing natural social behavior with our own kind."

I saw a hand-raised vulture used in a zoo education demonstration that was trained and cooperative in one place and wild in another. Inside his cage he was very territorial, and the keepers had to put on big gloves to remove him from his home. As soon as he saw a keeper approaching, his RAGE and FEAR systems turned on. Out in the public area, where he did flying demonstrations for the audience, his SEEKING system was turned on and he was a different bird. He would fly between two perches spaced about seventy-five feet apart on cue, and the big gloves were no longer necessary.

Using Novelty to Stimulate SEEKING

The simplest way to turn on the SEEKING system is to give animals something new to do. All humans and animals are interested in new things — animals like novelty so much they will sometimes cross a shock grid to get to a novel environment so they can explore it.[11] Some researchers think we can probably use an animal's liking for novelty as a measure of anhedonia, which is the loss of the ability to experience pleasure and a symptom of depression.[12] If they're right, that means you could use measures of an animal's interest in exploring novel objects and places as a measure of welfare.

However, you have to be careful not to force new things on animals (or on people). Animals like novelty if they can choose to investigate it; they fear novelty if you shove it in their faces. That's why the antelopes were so terrified of the yellow electrical sign, because they were being forced to walk past it. If the keepers had left them alone, the antelopes would have gone up to the sign and sniffed it.

The difference between forced and unforced novelty might also explain why animals get upset by tiny changes in their environments. Tiny changes are one kind of novelty. I wonder, if I went back and looked at all the times I've seen animals be terrified of a tiny detail, whether the real problem was that someone was trying to force them to approach or tolerate that new detail before they'd had a chance to explore it of their own free will.

SEEKING or Control?

The difference between forced and unforced novelty brings up questions about control. A lot of researchers and zookeepers believe that animals need to have control over their environments. Martin Seligman's studies of learned helplessness have been very influential on the subject of animals needing control, which keeps their SEEKING system turned on. In those studies two dogs were yoked together and given electric shocks. One of the animals could turn the shocks off; the other animal couldn't. Both dogs got the same amount of shock, but only one of the dogs could do something about it. Afterward, when the dogs were put in a new situation where they could escape a shock by jumping out of a compartment where they were being shocked, the dog that had been previously controlling the shocks immediately jumped out, but the one who could not control the shocks in the past didn't try to escape. The researchers believed that the yoked dog that had previously had no control over the shocks had learned to be helpless.[13] People working on welfare for captive animals concluded from the learned helplessness research that zoo animals need control in order to be happy.

I agree that zoo animals shouldn't be forced to do things they don't want to do. The use of any kind of force is usually aversive to animals (and people). But control isn't the best principle to base your enrichment programs on because "control" isn't a core emotion. Control is an aspect of the environment or of the animal's behavior in the environment, not a core system in the brain. If you made control the whole point of your enrichment program, you could end up creating barren environments where the animals have lots of control but are bored out of their minds. Daily chores are sometimes like that. People have a lot of control when they do their laundry, but you wouldn't want someone to put you in a zoo exhibit and give you piles of dirty laundry and a washer and dryer to keep you busy. That wouldn't be a very stimulating environment.

To understand how the learned helplessness experiments relate to enrichment, you have to go back to the brain. Escaping from an electric shock and not escaping from a shock are both behaviors, and all behaviors are driven by emotions. Very likely, the two animals were experiencing different emotions. The animal that could turn the shock off was probably feeling both FEAR of the shock and SEEKING a way to escape. The one that had no control would feel only FEAR. I think the learned helplessness researchers were right to believe that the animals that had no control over the shocks had become depressed. Their SEEKING system had been overinhibited by the uncontrollable shocks and they got depressed.[14]

For zoo animals, the negative emotion that is most frequently activated by lack of control is likely to be RAGE. We have to assume that some zoo animals may be living in a chronic state of restraint because they're locked up inside a zoo. You can make a zoo enclosure really nice and interesting and enriched, but it's still an enclosure. No wild animal evolved to be locked up inside a zoo or anywhere else.

Lack of control is bad for captive animals when it activates bad emotions. Most animals don't need constant control over their environment any more than most people need constant control. It's the emotion that matters, not the level of control an animal has.

PANIC and Companions

The good thing about using positive reinforcement is that the training program can satisfy some of the animal's social needs at the same time that it turns on SEEKING. When a trainer always uses positive reinforcement, the animal wants to spend time with him, and a trained animal will come eagerly running when it is time to practice with the blood pressure cuff and stethoscope. I have watched many of these sessions. All of the animals I've seen liked the treats they were fed, but it was obvious that some of them had developed a real social and emotional bond with their keepers. The animal and the human become attached to each other and their relationship is positive and warm.

That's important, because all animals need companions regardless of whether they're prey or predators, and making sure captive animals aren't lonely can be a big challenge for a zoo for practical reasons. If you've got two elephants and one elephant dies, you can't just go out and catch another one to keep the first one company. Capturing wild animals is horribly traumatic for the animals and we shouldn't do it.

The animal's normal level of sociability in the wild can also be a problem. In the wild, elephants live in extended families led by a matriarch. Most zoos do not have enough space to house an entire family of elephants. Today some zoos have adopted the policy that they will not acquire any new elephants unless they are old, retired animals that have been rescued from a circus or other place that keeps elephants.

Many primates can do well in zoos because family groups can be maintained and less space is required to house them. When a new chimpanzee exhibit was opened in Scotland, Jane Goodall remarked that chimpanzees might be better off in zoos because they would be safe from poaching and habitat loss.[15] Dr. Goodall said, "I would prefer that zoos did not exist but I choose to praise the ones that are doing the best job."[16]

Other big animals such as tigers, lions, and pandas don't need an extended family of animals to keep their PANIC emotion from being stirred up. However, no animal can live happily in isolation. The National Zoo in Washington, DC, had a poor situation with its male panda. He had a beautiful enclosure with new bamboo and other materials brought in every day for him to forage in. In the wild pandas spend most of their days foraging and eating bamboo, and the zoo was giving him lots of SEEKING opportunities. But socially he was a mess. He had been living with his girlfriend until the zoo took her away when she got pregnant and had a baby. Zoos often separate the males from the females after a baby is born because some male animals kill babies.

Without his girlfriend, the panda totally fell apart. He developed a weird mouth stereotypy where he was brushing his teeth over and over again with bamboo, and he would foam at the mouth. When his keeper came around, the panda wanted to play and get treats, and he loved doing medical demonstrations for the crowds because he was hungry for additional contact with his keeper. Pandas have always been thought to be solitary animals in the wild — although that may turn out to be wrong — so you'd expect him to be happy just foraging all day. But he was an animal who loved social contact. He was a photo-op bear who had been raised to have his picture taken and get treats from people in the zoo in China, and he loved the attention. He needed people to interact with.

I saw the bear when I was visiting Washington to give a talk at the Smithsonian, and I told the zoo that he needed to have interaction with his keeper for an hour or so every day. Either that or maybe he needed to go back to China and be a photo-op bear again.

In some cases, people can substitute for an animal's conspecifics if they have to.[17] This story about a parrot will give zookeepers insight into working with birds that pull out their feathers. Bird veterinarian Dr. Susan Orosz rescued a very neurotic African grey parrot from a pet shop. The parrot had been hand-reared by a breeder and trained to go on a person's finger so he was tamed to people. When she found him he was living in a barren cage where nobody paid any attention to him. He had pulled out his wing, tail, and chest feathers and was almost completely bald.

She gave the parrot a flock of human high school students in a Future Farmers of America class to substitute for the big flock the parrot would normally be living with if he were in the wild. The teacher put him in a big nice cage in the classroom covered with blankets on three sides to make him feel secure. Every night she would take him home in a dog carrier.

In the beginning the teacher gave the parrot five minutes on her finger in front of the classroom a few times a day. Within two weeks she was putting him on student fingers. She showed the students how to scratch the parrot on the head the way another parrot would, and after two weeks he started to approach the students himself.

They did this for three months, and at the end of that time his feather picking was down by half. Partly that was because the classroom was attached to a greenhouse where the students grew plants, and the parrot had learned that if he pulled out his wing and tail feathers he couldn't fly around the greenhouse. He made the connection between flying and having feathers because pulling out just a few of his large wing feathers made flying more difficult. By the end of the summer almost all his feathers had grown back and he didn't pull them out anymore.

Dr. Orosz chose an excellent, enriched environment for the parrot. The greenhouse reduced FEAR because the parrot could perch on trees under the foliage, which kept his instinctual fears of aerial predators from being turned on, and the students became his human flock so his PANIC system would not get turned on.

Last Thoughts

After *Animals in Translation* came out, I started getting calls from zoos that wanted me to consult on problems they were having with some of their animals. It was exciting to see the progress zoos have made. I have visited some beautiful rainforest exhibits where birds have lots of room to fly among the trees, and small primates have an abundance of plants to forage in, and all of the animals have places to hide and feel safe.

The big predators seem to be doing better, too, though we still don't know whether zoos will be able to give them everything they need to have a good life in captivity. Overall, researchers are finding that zoo enrichment programs are reducing stereotypies by 50 to 60 percent for all animals taken together.

The problem is that when an enrichment works, nobody knows which part of it worked. This makes it hard to improve on results. I believe that using the core emotion systems in the brain to analyze zoo enrichments will help us understand what is working and what isn't and create better lives for captive animals.

I believe that's true for all animals that live with people.

Afterword

Why Do I Still Work for the Industry?

I often get asked, "Why do you still work for the meat industry instead of being an activist against it?" A major factor that convinced me that I should continue to eat meat is that cattle and pigs during the 1970s, when I started my career, had good living conditions. The sows lived in pens and there were no sow stalls where the sow lived for most of her life where she could not turn around.

Handling and transport were atrocious in the 1970s but where the animals lived was decent. The beef cows were all out on pasture at family ranches, and the giant 20,000- to 60,000-head feedlots were dry and equipped with shades. During the first eleven years of my career in Arizona, I almost never saw disgusting, muddy feedlots because Arizona was super-hot with 100-degree temperatures and only six to eight inches of rainfall per year. All feedlots in Arizona had shades and the cattle grew really well in the dry climate. In my early years when I was starting my career, the beef and dairy cattle I worked with had nice places to live. I remember one dairy that had superb treatment of the animals by a manager who loved cows. When Michael Pollan, author of

The Omnivore's Dilemma, visited the feedlot where his steer lived, the muddy lot made a very bad impression on him.

Even in the 1970s there were a few people who were good managers and rough, abusive handling of cattle was not allowed. There were a few feedlots that had excellent handling and the local Swift plant where I started my career usually had good handling. From the very beginning of my career I saw that cattle could be raised right and given a good life and a painless death.

If my animal agriculture career had started with either egg-laying chickens or gross, muddy feedlots, my career might have taken a different direction. Chickens crammed in cages where they must sleep on top of each other would have turned me off. There are many people who became animal rights activists because one of their early animal experiences was absolutely awful. During the first five years of my career, I witnessed cattle handling that was shockingly bad, but a few progressive managers and stockpeople showed me that raising and handling cattle could be done with respect and kindness. One of those kind stockpeople was Allen, who worked vaccinating cattle for the big feedlots. He taught me how to operate the cattle squeeze chute very gently. He never got angry at the cattle; he just stayed calm. Then there was the Porters' beautiful Singing Valley Ranch. Penny and Bill Porter were the most wonderful, kind people and they were always gentle with their cattle. The Herefords they raised on their ranch were beautiful. These cattle had a wonderful life. This motivated me to work on improving the industry instead of working to convince people to stop eating meat. Knowing good, kind people who raised cattle had a huge influence on me. I knew the industry had its problems and it needed to be reformed, and these people made me believe it could be. At first I thought engineering could make all the improvements happen, but later in my career I learned that good engineering and design must be coupled with good management.

Do the Animals Know They Are Going to Die?

Often I get asked, "Do cattle know they are going to die?" While I was still in graduate school I had to answer this question. To find the answer I watched cattle go through the veterinary chute at a feedlot and then on the same day I watched them walk up the chute at the Swift plant. To my amazement, they behaved the same way in *both* places. If they knew that they were going to die, they should have acted wilder with more rearing and kicking at the Swift plant. At the plant, the handling was better and they were often calmer there.

Over the years I have done lots of thinking and have come to the conclusion that our relationship with the animals we use for food must be symbiotic. Symbiosis is a mutually beneficial relationship between two different living things. We provide the farm animals with food and housing and in return, most of the offspring from the breeding cows on the ranches are used for food. I vividly remember the day after I had installed the first center-track conveyor restrainer in a plant in Nebraska, when I stood on an overhead catwalk, overlooking vast herds of cattle in the stockyard below me. All these animals were going to their death in a system that I had designed. I started to cry and then a flash of insight came into my mind. None of the cattle that were at this slaughter plant would have been born if people had not bred and raised them. They would never have lived at all. People forget that nature can be very harsh, and death in the wild is often more painful and stressful than death in a modern plant. Out on a western ranch I saw a calf that had its hide ripped completely off on one side by coyotes. It was still alive and the rancher had to shoot it to put it out of its misery. If I had a choice, going to a well-run modern slaughter plant would be preferable to being ripped apart alive. I have lived in Colorado since 1990, and massive snowstorms have killed thousands of deer, elk, and cattle. Many animals died of starvation. Some ranchers could not get to their animals through twenty-foot snowdrifts, but the National Guard dropped hay to the cattle and the deer had to fend for themselves. Ranchers worked hard to save their animals. The natural environment can be a very tough world. Grazing animals such as cattle, sheep, and goats are a vital part of both sustainable and organic agriculture. Manure is used to fertilize the soil instead of chemicals. Grazing animals can also be used to improve pasture and help prevent land from turning into barren desert. Grazing is most beneficial in areas of the world with low rainfall. Allan Savory at the Center for Holistic Management explains that grazing has to mimic the behavioral patterns of herds of wildebeests or bison. The animals are stocked very tightly on a small piece of ground, then they move on. They mow the grass evenly, which helps plant diversity, and their manure fertilizes. Grazing done wrong will wreck pastures, but grazing done by herds of bison that constantly moved is what formed the Great Plains in the middle of America.

I am very concerned that programs around the world to convert grain into fuel will increase the intensification of animal agriculture. In both the United States and South America, prime pasture land is being converted to crops. In some areas, cattle are being moved off pastures and into feedlots. There is a lot of land where raising crops will increase soil erosion and damage the environment. Grazing animals is the best use for this land, and they help keep the land healthy.

Beef Slaughter Plant Tour

I have taken over a hundred nonindustry people through well-run beef plants that have restrainers and curved chute systems I have designed. Before they enter the plant, I let them watch trucks unload and cattle walk up the chute for about twenty minutes. They all expect the cattle to act crazy when they come off the trucks and they are amazed when the cattle stay calm. They just cannot believe that most cattle walk quietly into the plant without having to be prodded. I never rush this part of the tour because people cannot believe how calm the cattle can be until they see it. Of course this works only when all the distractions that scare cattle, such as shiny reflections or a hose on the floor, are removed from the system. After my visitors have watched a hundred cattle walk into the plant with only two animals out of the hundred making a sound, they start to get curious about what goes on behind the wall. My guests' SEEKING system is fully turned on, and then I let them walk through a door beside the cattle chute and watch the animals being shot with a captive bolt. The animals are instantly killed by a device that looks like a gigantic stainless-steel nail gun. The most common reaction is, "Oh, this is not as bad as I thought it would be."

Tours done wrong are disasters. I remember one lady who freaked out and almost fainted because she got a "bloody room tour." We were walking along the side of the building and had not even seen any cattle yet. Our tour guide suddenly opened a door and all she could see was blood all over everywhere, and the lady almost puked. At that point I took over the tour and took the lady to an overhead catwalk, which was over the cattle pens. From this vantage point, she could see the cattle walking into the building and all the pens of cattle. I told her that I wanted her to watch the cattle walk out of the pens and up the chute. The plant had one of my really nice curved systems and the cattle walked calmly. The handlers were calm and there was no yelling or whip cracking. We stood and watched for about fifteen minutes and she said, "I can't believe that they just walk quietly." Then I told her that it is like getting back on a horse and she should go back and open the door where she got so freaked out before. She walked down off the catwalk, opened the door, and said, "Not so bad now."

I often get asked, "How can you care about animals when you design slaughter plants?" Many people today are totally insulated from death, but every living thing eventually dies; this is the cycle of life. Since people are responsible for breeding and raising farm animals, they must also take the responsibility to give the animals living conditions that provide a decent life and a painless death. During the animal's life, both its physical needs and its emotional needs should be satisfied. Intensive farming systems need to be improved because the quality of the animals' lives is poor in some of them.

The more I observe and learn about how dogs are kept today, I am more convinced that many cattle have better lives than some of the pampered pets. Too many dogs are alone all day with no human or dog companions. Recently I walked down a residential street in a neighborhood close to my home, and I was appalled to hear three different dogs barking or whining in three different houses. Separation anxiety is a major problem for many dogs. One of the worst cases of separation anxiety was a dog who broke off its teeth trying to escape from a yard where he was alone all day. Just as this book was going to press, I visited Uruguay in South America. Pet dogs with collars were running around town with no leashes. Nobody was concerned about dog bites because the dogs were all well socialized. They were like the dog in Ted Kerasote's book, *Merle's Door.* After the hardcover edition of this book was published, I visited London and walked through a large park near the zoo. A lot of dogs were running around off leash with their owners nearby. This was a regular park, not a dog park. Families with little kids and the dogs were having a great time. It looked like the neighborhood of my youth.

Some people think death is the most terrible thing that can

happen to an animal. Dogs that run loose are often killed by cars, but their social life is probably better. Dogs that live a more confined existence are less likely to get killed, but their quality of life may be poorer unless their owners spend a lot of time playing and interacting with them. I think the most important thing for an animal is the *quality* of its life. A good life requires three things: health, freedom from pain and negative emotions, and lots of activities to turn on SEEKING and PLAY.

Challenging the Idea of Animal Emotions

Some people may not want to believe that animals really do have emotions. I think their own emotions are getting in the way of logic. When I read all the scientific evidence about electrical stimulation of subcortical brain systems, the only logical conclusion was that the basic emotion systems are similar in humans and all other mammals. I used cerebral, logical thinking to help reform slaughterhouses, and I used the same logical thought processes to fully accept the existence of emotions in animals.

Research Updates for the Paperback

A year after *Animals Make Us Human* was first published, scientific research continues to explore the existence of animal emotions. Recent dog research supports the extreme social awareness of dogs. Unlike mature wolves they understand what it means when a human points his finger at something. Further research supports the idea of a general purpose SEEKING system located in the nucleus accumbens.

Why has it taken researchers so long to scientifically recognize animal emotions when some of Panksepp's research is decades old? A big problem in science is the compartmentalization of data. Most of Panksepp's work was published in neuroscience literature. Neuroscientists have their journals, veterinarians have another set of journals, and animal behavior specialists have their own literature. Few people read literature outside their specific disciplines. If you would like to explore the latest research articles from neuroscience, veterinary, and animal behavior journals, try searching the databases on www.pubmed.com, www.scirus.com, and Google Scholar.[1]

Acknowledgments

This book would have been impossible without the help of my fabulous assistant, Cheryl Miller. When we got up against the final deadline, Cheryl gave up many weekends, making revisions, assembling the references, and putting the entire manuscript together. I also want to give my brilliant coauthor, Catherine Johnson, credit for coming up with the idea of linking the motivation of stereotypic behavior to Panksepp's four core emotions. When I read her first draft, I thought, "Wow, a breakthrough in thinking," and we both decided to use the concept throughout the book.

When I was young, I was a poor student and I had many mentors who helped me with my academic career. When I got my master's degree at Arizona State University, the animal science professors were not interested in my studies of cattle handling and different types of squeeze chutes. Fortunately, two creative professors who were outside the Animal Science Department were willing to serve on my committee and were really interested in my work. They were Foster Burton in the Construction Department and Mike Nielson in Industrial Design. Sometimes you have to go outside your field of study to find the right people. I wish to thank Stan Curtis in the Animal Science Department at the University of Illinois. He held open the back door, admitting me for my PhD even though my math ability was weak. Sometimes the academic system prevents good people from going into science. Einstein would be diagnosed with autism today and he was a patent clerk when he wrote his most important papers on relativity. I wonder what is happening to the Einsteins of today. Jaak Panksepp told me that some of his most brilliant students were thrown out of the Psychology Department due to poor grades in statistics. They were the ones who had the original ideas.

I also wish to thank Bill Greenough, my coadviser, for letting me do my project, and his neuroscience class opened up the world of brain research to me. I learned so much in his class and it was the best class I had for my PhD.

I would like to thank my editors. A special thank-you to Tina Pohlman at Harcourt, with whom we started the book. She liked our original ideas. We finished the book with Andrea Schulz when Tina moved to another publisher. Andrea really helped us to organize our thoughts into a coherent whole. I learned a lot from Andrea on how to organize my ideas. I wish to thank all the staff at Houghton Mifflin Harcourt who made this book possible.

TEMPLE GRANDIN

Beginning at the beginning, I am grateful to my agent, Elizabeth Kaplan, and to Temple's agent, Betsy Lerner, for their expert work not only in selling the book and negotiating the contracts, but also in refocusing the first proposal that Temple and I gave them to read. Betsy and Elizabeth helped us "find the basic principle," as Temple always says.

Our two editors at Harcourt, Tina Pohlman and her successor, Andrea Schulz, picked up from there, steering each chapter back to true whenever and wherever we'd gone astray. Thank you!

On the home front, three people kept the household going during the final months of wall-to-wall writing and revision: Martine, our children's caregiver; Christian Jackson, "big brother" to Chris, aide and friend to Jimmy and Andrew; and Ed Berenson, my husband, a man with the good nature, management skills, and sheer stamina to shoulder my share of the kids-and-house workload while also writing his own book and teaching at NYU. *Thank you.*

I would also like to thank a man I've never met: Jaak Panksepp, the author of *The Neuroscience of Emotion.* I've been studying Dr. Panksepp's book off and on for nearly ten years now, and I expect I'll be doing the same for the next ten. His work is rich, profound, and moving.

Finally, I would like to thank Temple for being Temple. Anyone reading this book knows that she has had a long and brilliant career in animal behavior and welfare, but less well-known is her importance to the world of autism. Temple has probably spent enough time talking to parents and young people with autism to fill up a second career, and she has a thriving sideline serving as a human subject in autism research too. We animals — human and nonhuman alike — are lucky to have her.

CATHERINE JOHNSON

Notes

1: What Do Animals Need?

1. Aubrey Manning and Marian Stamp Dawkins, *An Introduction to Animal Behavior*, 5th ed. (Cambridge: Cambridge University Press, 1998), 371–72.
2. Christoph Wiedenmayer, "Causation of the Ontogenetic Development of Stereotypic Digging in Gerbils," *Animal Behaviour* 53, no. 3 (1997): 461–70.
3. Karen Overall, *Clinical Behavioral Medicine for Small Animals* (St. Louis: Mosby, 1997). There are hundreds and hundreds of journal articles where the neurotransmitter systems in the brains of mammals have been studied. Addictive drugs have similar effects on both people and animals. You can search the literature and obtain free summaries of scientific studies by going to the PubMed database, which searches the National Library of Medicine. Type *pubmed* into Google or some other search engine.
4. Temple Grandin, *Thinking in Pictures* (New York: Vintage Press, 1995). Describes the author's experiences with dissecting a human brain and a pig's brain.
5. Jaak Panksepp, *Affective Neuroscience: The Foundations of Human and Animal Emotions* (New York: Oxford University Press, 1998).
6. Jaak Panksepp, "Affective Consciousness: Core Emotional Feelings in Animals and Humans," *Consciousness and Cognition* 14 (2005): 30–80. Dopamine is the neurotransmitter that operates the SEEKING system. Brain circuits that contain the neurotransmitter dopamine are most responsive to the anticipation of a reward. Dopamine levels are more likely to rise when there are unexpected rewards or positive novel stimuli. See Panksepp, "Affective Consciousness: Core Emotional Feelings in Animals and Humans," 47. S.P.D. Judd and T. S. Collett, "Multiple Stored Views and Landmark Guidance in Ants," *Nature* 392 (1998): 710–14. I include this reference to support my hypothesis that the orienting response, where an animal turns

toward a novel stimulus, is the most basic form of SEEKING. When an animal pauses after orienting, I think it is making a "same or different" comparison between the thing it is seeing or hearing and the information stored in memory. Ants return from foraging trips and compare new images with stored images to find their way back home. On this most primitive level, the nervous system determines same versus different, and the ant stops and "orients" to make a decision. N. Bunzeck and E. Duzel, "Absolute Coding of Stimulus Novelty in the Human Substantia Nigra/VTA," *Neuron* 51 (2006): 369–79; Jeffrey Burgdorf and Jaak Panksepp, "The Neurobiology of Positive Emotions," *Neuroscience and Biobehavioral Reviews* 30 (2006): 173–87. These papers review research on subcortical systems of seeking and other positive emotions in animals; Jaak Panksepp, "On the Embodied Neural Nature of Core Emotional Affects," *Journal of Consciousness Studies* 12, no. 8 (2005): 158–84. This paper explains how the positive emotions of PLAY, LUST, and CARE need a general-purpose SEEKING system that helps an animal anticipate a reward. J. F. Lisman and A. A. Grace, "The Hippocampal-VTA Loop Controlling Entry into Long-Term Memory," *Neuron* 46 (2005): 703–13. The hippocampus and the ventral tegmental area contain circuits that detect novelty and turn on the dopamine system. This would be part of the SEEKING circuits.

7. James Olds, *Drives and Reinforcements: Behavioral Studies of Hypothalamic Function* (New York: Raven Press, 1977).
8. Kent C. Berridge and Terry E. Robinson, "Parsing Reward," *Trends in Neuroscience* 26, no. 9 (2003): 507–13.
9. Jaak Panksepp, "On the Embodied Neural Nature of Core Emotional Affects," *Journal of Consciousness Studies* 12, no. 8 (2005): 177.
10. Alexis Faure, Sheila M. Reynolds, Jocelyn M. Richard, and Kent C. Burridge, "Mesolimbic Dopamine in Desire and Dread: Enabling Motivation to Be Generated by Localized Glutamate Disruption of the Nucleus Acumbens," *Journal of Neuroscience* 28, no. 28 (2008): 7184–7192. Stimulating the entire "pleasure center" with an electrode turns on SEEKING, and the animal approaches. In this study, more precise methods were used to stimulate different locations in the nucleus acumbens. When one location was stimulated, the rat engaged in appetitive behaviors, and stimulation in another location caused the rat to engage in fearful behavior. Amy Maxman, "Dopamine's Role Linked to Emotions," *Science News*, August 2, 2008, p. 8. Richard Palmiter at the University of Washington explains that in both cases, the brain's dopamine system in the nucleus acumbens is saying, "Hey, pay attention to your environment." I think this is the basic "orienting" part of SEEKING. This newly discovered brain mechanism prevents the animal from getting eaten by a predator. My theory is that having connections in the "pleasure center" to the other negative core emotions

instantly turns off the positive emotion when the animal is in danger. A delicate balance between several competing neurotransmitters determines whether or not the animal will SEEK something pleasurable or shut down the SEEKING system and activate a negative emotion to help it survive.

11. Jaak Panksepp, "Aggression Elicited by Electrical Stimulation of the Hypothalamus in Albino Rats," *Physiology and Behavior* 6, no. 4 (1971): 321–29; W. R. Hess, *The Functional Organization of the Diencephalon* (New York: Grune and Stratton, 1957); Allan Siegel, *The Neurobiology of Aggression and Rage* (Boca Raton, FL: CRC Press, 2005).
12. Joseph LeDoux, *The Emotional Brain* (New York: Simon and Schuster, 2006).
13. Temple Grandin, "Assessment of Stress during Handling and Transport," *Journal of Animal Science* 75 (1997): 249–57. Reviews studies on the FEAR system in the brain.
14. J. P. Semitelou, J. K. Yakinthos, and C. Sue Carter, "Neuroendocrine Perspectives on Social Attachment and Love," *Psychoneuroendocrinology* 23, no. 8 (1998): 779–818; Panksepp, "Affective Consciousness: Core Emotional Feelings in Animals and Humans," 55. Oxytocin is more effective than opioids for reducing stress caused by separation from the mother.
15. J. Panksepp, L. Normansell, J. F. Cox, and S. Siviy, "Effects of Neonatal Decortication on the Social Play of Juvenile Rats," *Physiology and Behavior* 56, no. 3 (1994): 429–43: Play is a basic emotion that originates from subcortical brain systems. Animals with no cortex will still play. Michael Okun, Dawn Bowers, Utaka Springer, Nathan Shapira, Donald Malone, Ali Rezai, Bart Nuttin, Kenneth Heilman, Robert Morecraft, Steven Rasmussen, Benjamin Greenberg, Kelly Foote, and Wayne Goodman, "What's in a 'Smile?' Intra-operative Observations of Contralateral Smiles Induced by Deep Brain Stimulation," *Neurocase* 10, no. 4 (2004): 271–79: Electrical stimulation of the nucleus accumbens in humans causes people to smile. Jaak Panksepp and Jeff Burgdorf, "'Laughing' Rats and the Evolutionary Antecedents of Human Joy?" *Physiology and Behavior* 79, no. 3 (2003): 533–47: Stimulating a region of the rat brain that is analogous to the nucleus accumbens in people elicits a playful chirping sound.
16. M. C. Diamond, D. Krech, and M. R. Rosenzweig, "The Effects of an Enriched Environment on the Histology of the Rat Cerebral Cortex," *Journal of Comparative Neurology* 123, no. 1 (1964): 111–20.
17. M. C. Diamond, F. Law, H. Rhodes, B. Lindner, M. R. Rosenzweig, D. Krech, and E. L. Bennett, "Increases in Cortical Depth and Glia Numbers in Rats Subjected to Enriched Environment," *Journal of Comparative Neurology* 128, no. 1 (1966): 117–25.
18. Fred R. Volkmar and William T. Greenough, "Rearing Complexity Affects Branching of Dendrites in the Visual Cortex of the Rat," *Science* 176 (1972): 1445–47.

19. Temple Grandin, "Effect of Rearing Environment and Environmental Enrichment on Behavior and Neural Development of Young Pigs" (doctoral dissertation, University of Illinois, Urbana, Illinois, 1989).
20. Jeffrey Rushen and Georgia Mason, "A Decade-or-More's Progress in Understanding Stereotypic Behavior," in *Stereotypic Animal Behaviour*, ed. Jeffrey Rushen and Georgia Mason (Wallingford, UK: CABI Publishing, 2006), 1–18.
21. Steffen Werner Hansen and Leif Lau Jeppesen, "Temperament, Stereotypies and Anticipatory Behavior as Measures of Welfare in Mink," *Applied Animal Behaviour Science* 99, no. 1–2 (2006): 172–82.
22. Georgia Mason, "Are Wild Caught Animals 'Protected' from Stereotypy When Placed in Captivity?" in *Stereotypic Animal Behaviour*, ed. Rushen and Mason, 196, Box 7.1.
23. Antonio R. Damasio, Thomas J. Grabowski, Antoine Bechara, Hanna Damasio, Laura L. B. Ponto, Josef Parvizi, and Richard D. Hichwa, "Subcortical and Cortical Brain Activity during the Feeling of Self-Generated Emotions," *Nature Neuroscience* 3, no. 10 (2000): 1049–56. The work of Dr. Damasio shows how the emotional system works. His work shows how the emotions are located in the subcortical parts of the brain. From these studies, I conclude that the basic emotions in humans and other mammals are similar, because the subcortical regions of the brain are similar. I have always told my students that to understand behavior, you have to relate the observable behavior back to brain function. Emotional complexity will increase as the size of the neocortex is increased, but the basic core emotional system will remain the same. As a person with autism, I have simpler, less complex emotions because some large circuits to my frontal cortex are missing, but my basic core emotions of fear, rage, panic, and seeking are strong.

2: A Dog's Life

1. Brian Hare and Michael Tomasello, "Human-like Social Skills in Dogs?" *Trends in Cognitive Sciences* 9, no. 9 (September 2005): 439–44.
2. Juliane Bräuer, Juliane Kaminski, Julia Riedel, Josep Call, and Michael Tomasello, "Making Inferences about the Location of Hidden Food: Social Dog, Causal Ape," *Journal of Comparative Psychology* 120, no. 1 (2006): 38–47.
3. A *reinforcer* is anything that follows a behavior and makes the behavior more likely to happen. I'll talk about that a lot more later on in the chapter.
4. Robert K. Wayne, "Molecular Evolution of the Dog Family," www.idir.net/~wolf2dog/wayne2.htm: "The domestic dog is an extremely close relative of the gray wolf, differing from it by at most 0.2% of mtDNA sequence." The Wolfdogs Resource: www.idir.net/~wolf2dog/index.html.

5. Kerstin Lindblad-Toh et al., "Genome Sequence, Comparative Analysis and Haplotype Structure of the Domestic Dog," *Nature* 438, no. 7069 (2005): 803–19.
6. L. David Mech, "Alpha Status, Dominance, and Division of Labor in Wolf Packs," *Canadian Journal of Zoology* 77, no. 8 (1999): 1196–1203.
7. Or the way people lived in traditional societies.
8. Mech, "Alpha Status, Dominance, and Division of Labor in Wolf Packs."
9. Adolph Murie, *The Wolves of Mount McKinley,* Fauna Series No. 5 (Washington, DC: U.S. National Park Service, 1944).
10. The fact that three females gave birth in a wolf pack — even a forced pack put together by humans — is more evidence that wolf packs don't have an alpha female who prevents the subordinate females from mating.
11. Mike Stark, "Druid Wolf Pack Dwindles in Park," *Billings Gazette,* February 22, 2003, www.billingsgazette.com/newdex.php?display=rednews/2003/02/22/build/wyoming/30-wolf-pack.inc.
12. Cesar Millan with Melissa Jo Peltier, *Cesar's Way: The Natural, Everyday Guide to Understanding and Correcting Common Dog Problems* (New York: Harmony Books, 2006), 114, 146.
13. Millan, *Cesar's Way,* 24–25.
14. I don't know what a free-living family of *dogs* would be like because a grown dog is psychologically immature compared to a grown wolf. I'll talk about that later.
15. No one knows how feral dogs are organized socially. They usually travel in groups, but whether those groups are packs of unrelated individual dogs or families, we don't know.
16. Millan, *Cesar's Way,* 236.
17. Researchers have found dog remains buried under huts going back fourteen thousand years, which is still before agriculture was invented. At some point individual dogs started to be associated with individual humans or human families.
18. Half of Cesar's dogs live at the Dog Psychology Center permanently. That's their home. The other half are temporary boarders. Some of the temporaries come from shelters that are going to euthanize them if they can't be socialized, and the rest are healthy dogs that stay at the center when their owners travel. Millan, *Cesar's Way,* 9.
19. Deborah Goodwin, John W. S. Bradshaw, and Stephen M. Wickens, "Paedomorphosis Affects Agonistic Visual Signals of Domestic Dogs," *Animal Behaviour* 53, no. 2 (1997): 297–304.
20. Ibid., 302.
21. Karen B. London and Patricia B. McConnell, *Feeling Outnumbered? How to Manage and Enjoy Your Multi-Dog Household* (Black Earth, WI: Dog's Best Friend, Ltd., 2001), 1.

22. Dorit Urd Feddersen-Petersen, "Social Behaviour of Dogs and Related Canids," in *The Behavioural Biology of Dogs*, ed. Per Jensen (Wallingford, UK: CABI International, 2007), 105–19.
23. Six breeds showing one or no submissive signals: Cavalier King Charles spaniel (0), Norfolk terrier (1), Shetland sheepdog (1), French bulldog (1), cocker spaniel (1), large Munsterlander (a hunting dog related to the German longhaired pointer) (1). Goodwin, Bradshaw, and Wickens, "Paedomorphosis Affects Agonistic Visual Signals of Domestic Dogs," 297.
24. Feddersen-Petersen, "Social Behaviour of Dogs and Related Canids," 115.
25. Peter L. Borchelt, Randall Lockwood, Alan M. Beck, and Victoria L. Voith, "Attacks by Packs of Dogs Involving Predation on Human Beings," *Public Health Reports* 98, no. 1 (January–February 1983): 58.
26. Feddersen-Petersen, "Social Behaviour of Dogs and Related Canids," 113–14.
27. Michael Milstein, "Wolf Pack's Activities Like a Soap Opera," *Billings Gazette*, May 28, 2000.
28. Ted Kerasote, *Merle's Door: Lessons from a Freethinking Dog* (Orlando, FL: Harcourt, 2007).
29. M. Falcon, "Frasier's Dog Eddie Barks against Bites" (Spotlight on Health), *USA Today*, June 4, 2001.
30. Anna-Elisa Liinamo, Linda van den Berg, Peter A. J. Leegwater, Matthijs B. H. Schilder, Johan A. M. van Arendonk, and Bernard A. van Oost, "Genetic Variation in Aggression-Related Traits in Golden Retriever Dogs," *Applied Animal Behaviour Science* 104, no. 1–2 (2007): 104.
31. Peter Aldhous, "Pampered Pooches Are Canine Peter Pans," *New Scientist* 2075 (March 29, 1997): 5.
32. Patricia B. McConnell, *For the Love of a Dog: Understanding Emotion in You and Your Best Friend* (New York: Ballantine Books, 2005), 85.
33. Behavioral medicine means veterinarians learning to diagnose and treat behavior problems systematically, both through training and through physical treatments including medication. It's a little like psychiatry for animals.
34. Overall, *Clinical Behavioral Medicine for Small Animals*, 11.
35. Panksepp, *Affective Neuroscience*, 190.
36. McConnell, *For the Love of a Dog*, 194.
37. Ibid., 189.
38. Ibid., 190.
39. All Labs have a very high threshold for pain, which is good for families with children. Step on a mutt's paw and he'll yelp in pain; step on a Lab's paw and he'll barely notice it. Since some animals can bite when they experience sudden pain, the fact that Labs don't experience much pain makes them safer.
40. The other recognized forms of aggression are territorial, maternal, food-related, interdog, predatory, possessive, idiopathic, pain, and play aggression.

Dominance and fear aggression are the most common types of problem aggression that cause people to get help handling their dogs. Overall, *Clinical Behavioral Medicine for Small Animals*, 102.

41. Ned H. Kalin, Steven E. Shelton, Richard J. Davidson, and Ann E. Kelley, "The Primate Amygdala Mediates Acute Fear but Not the Behavioral and Physiological Components of Anxious Temperament," *The Journal of Neuroscience* 21, no. 6 (March 15, 2001): 2067–74.
42. Overall, *Clinical Behavioral Medicine for Small Animals*, 109, 124.
43. On male dogs, the bandage is never wrapped over their privates.
44. Nancy G. Williams and Peter L. Borchelt, "Full Body Restraint and Rapid Stimulus Exposure as a Treatment for Dogs with Defensive Aggressive Behavior: Three Case Studies," *International Journal of Comparative Psychology* 16, no. 4 (2003): 226–36.
45. McConnell, *For the Love of a Dog*, 208.
46. Nicola J. Rooney and John W. S. Bradshaw, "An Experimental Study of the Effects of Play upon the Dog-Human Relationship," *Applied Animal Behaviour Science* 75, no. 2 (2002): 161–76.
47. McConnell, *For the Love of a Dog*, 176–77.
48. Ibid., 124–25.
49. Jerome Groopman, "Dog Genes: Medical Dispatch Pet Scans," *New Yorker* (May 10, 1999), 46.
50. Overall, *Clinical Behavioral Medicine for Small Animals*, 120.
51. When a brain cell releases an excitatory neurotransmitter, the "downstream" neurons become "excited" and they fire, too. Inhibitory neurotransmitters do the opposite.
52. McConnell, *For the Love of a Dog*, 186–87.
53. Ibid., 224.
54. Coppola C. L., T. Grandin, and R. M. Enns, "Human Interaction and Cortisol: Can Human Contact Reduce Stress in Shelter Dogs?" *Physiology and Behavior* 87 (2006): 537–41; Morrell, M., "Going to the Dogs," *Science* 325 (2009): 1062–65; Topal, J., G. Gergely, A. Erdohegyi, G. Csibra, and A. Miklosi, "Differential Sensitivity to Human Communication in Dogs, Wolves, and Human Infants," *Science* 325 (2009): 1269–71.
55. You can always carry the leash with you in case you need to put it on your dog to cross a busy street or when other dog owners ask you to.

3: Cats

1. Overall, *Clinical Behavioral Medicine for Small Animals*, 45.
2. Ibid.
3. James A. Serpell, "Domestication and History of the Cat," in *The Domestic*

Cat: The Biology of Its Behaviour, 2nd ed., ed. Dennis C. Turner and Patrick Bateson (Cambridge: Cambridge University Press, 2000), 180–92.

4. Dieter C. T. Kruska, "On the Evolutionary Significance of Encephalization in Some Eutherian Mammals: Effects of Adaptive Radiation, Domestication, and Feralization," *Brain, Behavior and Evolution* 65, no. 2 (2005): 73–108.
5. John Bradshaw and Charlotte Cameron-Beaumont, "The Signalling Repertoire of the Domestic Cat and Its Undomesticated Relatives," in *The Domestic Cat*, ed. Turner and Bateson, 88.
6. Karen Pryor, *Don't Shoot the Dog!: The New Art of Teaching and Training* (New York: Bantam Books, 1999), 113.
7. Nicholas Dodman, *The Cat Who Cried for Help: Attitudes, Emotions, and the Psychology of Cats* (New York: Bantam Books, 1997), 235.
8. Overall, *Clinical Behavioral Medicine for Small Animals*, 75.
9. Bradshaw and Cameron-Beaumont, "The Signalling Repertoire of the Domestic Cat and Its Undomesticated Relatives," 72.
10. Dodman, *The Cat Who Cried for Help*, 47–49.
11. A lot of autistic people have problems with facial recognition. I've had parents tell me that their young autistic child mistook a complete stranger for them based on things like the stranger having a similar hairstyle to the parent.
12. A wild predator that's losing its fear of humans and starting to come into populated areas is dangerous. I'll talk about that in my chapter on wildlife.
13. The word for that is *piloerection*, which means erection of the hairs on the skin. When you say something makes your hair stand on end, that's piloerection.
14. Dennis C. Turner, "The Human-Cat Relationship," in *The Domestic Cat*, ed. Turner and Bateson, 198.
15. Michael Mendl and Robert Harcourt, "Individuality in the Domestic Cat: Origins, Development, and Stability," in *The Domestic Cat*, ed. Turner and Bateson, 193–206.
16. Turner, "The Human-Cat Relationship," 202–3.
17. Overall, *Clinical Behavioral Medicine for Small Animals*, 54.
18. Ibid., 53–54.
19. Dodman, *The Cat Who Cried for Help*, 109.
20. Ibid., 112.
21. Ibid.
22. Overall, *Clinical Behavioral Medicine for Small Animals*, 160.
23. Dodman, *The Cat Who Cried for Help*, 116.
24. Overall, *Clinical Behavioral Medicine for Small Animals*, 160.
25. Panksepp, *Affective Neuroscience*, 161.
26. Ibid., 212.

27. Elkhonon Goldberg, *The Executive Brain: Frontal Lobes and the Civilized Mind* (New York: Oxford University Press, 2001).
28. Joaquin M. Fuster, *The Prefrontal Cortex*, 3rd ed. (Philadelphia: Lippincott-Raven, 1997).
29. Dodman, *The Cat Who Cried for Help*, 52.
30. Overall, *Clinical Behavioral Medicine for Small Animals*, 168.
31. Predator animals have a natural instinct to hunt, but they have to be taught by their parents that particular animals are prey. However, that doesn't mean it's safe to have loose rabbits around cats. Cats are super-predators, and rapid movement triggers their *prey chase drive.* A cat could kill a pet rabbit by instinct without realizing the rabbit is good to eat.
32. B. Mike Fitzgerald and Dennis C. Turner, "Hunting Behaviour of Domestic Cats and Their Impact on Prey Populations," in *The Domestic Cat*, ed. Turner and Bateson, 155.
33. Patrick Bateson, "Behavioural Development in the Cat" in *The Domestic Cat*, ed. Turner and Bateson, 15.
34. Matt Jarvis, Julia Russell, and Phil Gorman, *Angles on Psychology*, 2nd ed. (Cheltenham, UK: Nelson Thornes, 2004).
35. Interview with Karen L. Overall by Marcella Durand, retrieved from www.catsplay.com/thedailycat/2003-03-31/culture_interview/int_koverall/int_koverall.html on November 21, 2007.
36. Ibid.
37. Anything that makes a noise will work. Dolphin trainers use a whistle.
38. Miranda Hersey Helin, "Clicker Training Cats as Service Animals. . . ?," retrieved from www.clickertraining.com/node/1418 on May 13, 2008.
39. Karen Pryor, "Charging the Clicker," retrieved from www.clickertraining.com/node/824 on May 13, 2008.
40. Robin Rockey, "A Different Approach," *Cat Fancy* (October 2008): 12–13.

4: Horses

1. "Horses vs. Vehicles . . . Your Safety," retrieved from www.chatsworthecho.org/vehicle_code.html on May 14, 2008.
2. Xenophon, "On Horsemanship," Section XI, retrieved from www.classicreader.com/read.php/bookid.1807/sec.11/ on May 15, 2008.
3. Natalie K. Waran and Rachel Casey, "Horse Training," in *The Domestic Horse: The Evolution, Development and Management of Its Behaviour*, ed. Daniel Mills and Sue McDonnell (Cambridge: Cambridge University Press, 2005), 193.
4. S. Henry, D. Hemery, M. A. Richard, and M. Hausberger, "Human–Mare Relationships and Behaviour of Foals toward Humans," *Applied Animal Behaviour Science* 93, no. 3–4 (2005): 341.

5. Ibid., 341–62.
6. J. Lanier and T. Grandin, "The Relationship between *Bos taurus* Feedlot Cattle Temperament and Cannon Bone Measurements," *Proceedings, Western Section, American Society of Animal Science* 53 (2002): 97–101.
7. Janne Winther Christensen, Linda Jane Keeling, and Birte Lindstrøm Nielsen, "Responses of Horses to Novel Visual, Olfactory and Auditory Stimuli," *Applied Animal Behaviour Science* 93, no. 1–2 (2005): 53–65.
8. Evelyn B. Hanggi, "The Thinking Horse: Cognition and Perception," *Proceedings of the 51st Annual Convention of the American Association of Equine Practitioners* (2005): 251.
9. When a horse is crosstied, he has two ropes attached to his halter with each end tied to opposite walls of the barn alley, which keeps him in the middle of the alley. That lets people work on both sides of the horse to groom or shoe him.
10. Lynn M. McAfee, Daniel S. Mills, and Jonathan J. Cooper, "The Use of Mirrors for the Control of Stereotypic Weaving Behaviour in the Stabled Horse," *Applied Animal Behaviour Science* 78, no. 2 (2002): 159–73.
11. Jonathan J. Cooper, Lisa McDonald, and Daniel S. Mills, "The Effect of Increasing Visual Horizons on Stereotypic Weaving: Implications for the Social Housing of Stabled Horses," *Applied Animal Behaviour Science* 69, no. 1 (2000): 67–83.
12. E. Søndergaard and J. Ladewig, "Group Housing Exerts a Positive Effect on the Behaviour of Young Horses during Training," *Applied Animal Behaviour Science* 87, no. 1–2 (2004): 105–18.
13. Jared Diamond, *Guns, Germs, and Steel: The Fates of Human Societies* (New York: W. W. Norton, 1997), 162.
14. Ibid.
15. Ibid.
16. J. R. Silver, "Spinal Injuries Resulting from Horse Riding Accidents," *Spinal Cord* 40, no. 6 (2002): 264–71.
17. Paul McGreevy, *Equine Behavior: A Guide for Veterinarians and Equine Scientists* (Philadelphia: W. B. Saunders, 2004), 294–96.
18. K. McGee, J. Lanier, and T. Grandin, "Characteristics of Horses Sold at Auctions and in Slaughter Plants" (survey done for the U.S. Department of Agriculture, 2001).
19. Xenophon, "On Horsemanship," trans. H. G. Dakyns, Gutenberg Project, www.gutenberg.org/dirs/etext98/hrsmn10.txt.
20. Monty Roberts, *Horse Sense for People: The Man Who Listens to Horses Talks to People* (New York: Penguin, 2000).
21. Tom Dorrance, *True Unity: Willing Communication between Horse and Human* (Bruneau, ID: Give-It-A-Go Enterprises, 1987), 9.
22. McGreevy, *Equine Behavior*, 296.

23. C. J. Nicol, "Equine Learning: Progress and Suggestions for Future Research," *Applied Animal Behaviour Science* 78, no. 2 (2002): 193–208.
24. Ibid., 197.
25. Negative reinforcement could also work by activating the RAGE system, or by activating some combination of FEAR and RAGE. Since horses are high-fear animals, I assume that negative reinforcement usually activates their FEAR system.
26. Karen Pryor, "Samples of Negative Reinforcement," www.clickertraining.com/node/274.
27. J. L. Williams, T. H. Friend, C. H. Nevill, and G. Archer, "The Efficacy of a Secondary Reinforcer (Clicker) during Acquisition and Extinction of an Operant Task in Horses," *Applied Animal Behaviour Science* 88, no. 3–4 (2004): 331–41.
28. Ibid.
29. Alexandra Kurland, *Clicker Training for Your Horse* (Waltham, MA: Sunshine Books, 2007).
30. Miranda Hersey Helin, "Expo Faculty Profile: Alexandra Kurland, Horse Trainer," www.clickertraining.com/node/79.
31. William C. Follette, Peter J. N. Linnerooth, and L. E. Ruckstuhl Jr., "Positive Psychology: A Clinical Behavior Analytic Perspective," *Journal of Humanistic Psychology* 41, no. 1 (Winter 2001): 118–19.
32. Ibid.
33. Ibid.
34. "History of Clicker Training I," Karen Pryor's Acceptance Speech, Annual Award for Excellence in the Media Association for Behavior Analysis, Chicago, May 23, 1997, www.clickertraining.com/node/153.
35. McGreevy, *Equine Behavior*, 102.
36. Dawnery L. Ferguson and Jesus Rosales-Ruiz, "Loading the Problem Loader: The Effects of Target Training and Shaping on Trailer-Loading Behavior of Horses," *Journal of Applied Behavior Analysis* 34, no. 4 (2001): 411.
37. Ibid., 421.
38. Helin, "Expo Faculty Profile: Alexandra Kurland, Horse Trainer."

5: Cows

1. Radka Šárová, Marek Špinka, and José L. Arias Panamá, "Synchronization and Leadership in Switches between Resting and Activity in a Beef Cattle Herd — A Case Study," *Applied Animal Behaviour Science* 108, no. 3–4 (2007): 327–31.
2. David Val-Laillet, Anne Marie de Passillé, Jeffrey Rushen, and Marina A. G. von Keyserlingk, "The Concept of Social Dominance and the Social Distri-

bution of Feeding-Related Displacements between Cows," *Applied Animal Behaviour Science* 111, no. 1–2 (2008): 158–72.

3. U.S. Department of Agriculture Economic Research Service, "U.S. Beef and Cattle Industry: Background Statistics and Information," www.ers.usda .gov/news/BSECoverage.htm.
4. Paul Hemsworth, "Human-Livestock Interaction," in *The Well-Being of Farm Animals: Challenges and Solutions,* ed. G. John Benson and Bernard E. Rollin (Ames, IA: Blackwell Publishing Professional, 2004), 22.
5. D. F. Waynert, J. M. Stookey, K. S. Schwartzkopf-Genswein, J. M. Watts, and C. S. Waltz, "The Response of Beef Cattle to Noise during Handling," *Applied Animal Behaviour Science* 62, no. 1 (1999): 27–42.
6. All animals and people are curiously afraid, but prey animals might be more afraid than predator animals.
7. Peter M. Milner, *The Autonomous Brain* (Mahwah, New Jersey: Lawrence Erlbaum Associates, 1999).
8. Cyril Herry, Dominik R. Bach, Fabrizio Esposito, Franceso Di Salle, Walter J. Perrig, Klaus Scheffler, Andreas Lüthi, and Erich Seifritz, "Processing of Temporal Unpredictability in Human and Animal Amygdala," *Journal of Neuroscience* 27, no. 22 (2007): 5958–66.
9. Paul J. Whalen, "The Uncertainty of It All," *Trends in Cognitive Sciences* 11, no. 12 (2007): 499.
10. Ibid.
11. Temple Grandin and Mark J. Deesing, "Behavioral Genetics and Animal Science," in *Genetics and the Behavior of Domestic Animals,* ed. Temple Grandin (San Diego: Academic Press, 1998), 1–30.
12. Temple Grandin, "Factors That Impede Animal Movement in Slaughter Plants," *Journal of the American Veterinary Medical Association* 209, no. 4 (1996): 757–59.
13. Hemsworth, "Human-Livestock Interaction," 22.
14. Temple Grandin, "Voluntary Acceptance of Restraint by Sheep," *Applied Animal Behaviour Science* 23, no. 3 (1989): 257–61.
15. Steve Cote, *Stockmanship: A Powerful Tool for Grazing Land Management* (Arco, ID: U.S. Department of Agriculture, Natural Resources Conservation Services, 2004).
16. Ibid., 3.
17. Ibid., 119.
18. Caroline Lee, Kishore Prayaga, Matt Reed, and John Henshall, "Methods of Training Cattle to Avoid a Location Using Electrical Cues," *Applied Animal Behaviour Science* 108, no. 3–4 (2007): 229–38.
19. A. R. Tiedemann, T. M. Quigley, L. D. White, W. S. Lauritzen, J. W. Thomas, and M. L. McInnis, "Electronic (Fenceless) Control of Livestock,"

Research Paper PNW-RP-510 (Portland, OR: U.S. Department of Agriculture, 1999).

20. George W. Rankin, *The Life of William Dempster Hoard* (Fort Atkinson, WI: W. D. Hoard and Sons Press, 1925).
21. Paul H. Hemsworth and Grahame J. Coleman, *Human-Livestock Interactions: The Stock Person and the Productivity and Welfare of Intensely Farmed Animals* (Wallingford, UK: CABI International, 1998).
22. Panksepp, "Affective Consciousness: Core Emotional Feelings in Animals and Humans," 53.
23. R. Müller, K. S. Schwartzkopf-Genswein, M. A. Shah, and M. A. G. von Keyserlingk, "Effect of Neck Injection and Handler Visibility on Behavioral Reactivity of Beef Steers," *Journal of Animal Science* 86, no. 5 (2008): 1215–22.
24. Temple Grandin with Mark Deesing, *Humane Livestock Handling* (North Adams, MA: Storey Publishing, 2008), 63.
25. Joseph M. Stookey and J. M. Watts, "Low Stress Restraint, Handling, and Weaning of Cattle," in *Livestock Handling and Transport,* ed. Temple Grandin (Wallingford, UK: CABI International, 2007), 65–75.
26. D. B. Haley, D. W. Bailey, and J. M. Stookey, "The Effects of Weaning Beef Calves in Two Stages on Their Behavior and Growth Rate," *Journal of Animal Science* 83 (2005): 2205–14.
27. Joseph M. Stookey and Jon M. Watts, "Production Practices and Well-Being: Beef Cattle," in *The Well-Being of Farm Animals,* ed. Benson and Rollin, 194.
28. Knut Egil Bøe and Gry Færevik, "Grouping and Social Preferences in Calves, Heifers and Cows," *Applied Animal Behaviour Science* 80, no. 3 (2003): 175–80.
29. Gry Færevik, Inger L. Andersen, Margit B. Jensen, and Knut E. Bøe, "Increased Group Size Reduces Conflicts and Strengthens the Preference for Familiar Group Mates after Regrouping of Weaned Dairy Calves (*Bos taurus*)," *Applied Animal Behaviour Science* 108, no. 3–4 (2007): 215–28.
30. Stookey and Watts, "Production Practices and Well-Being: Beef Cattle," in *The Well-Being of Farm Animals,* ed. Benson and Rollin, 201.
31. Satu Raussi, Alain Boissy, Eric Delval, Philippe Pradel, Jutta Kaihilahti, and Isabelle Veissier, "Does Repeated Regrouping Alter the Social Behaviour of Heifers?" *Applied Animal Behaviour Science* 93, no. 1–2 (2005): 1–12.
32. Ibid.
33. Ibid.
34. Stookey and Watts, "Production Practices and Well-Being: Beef Cattle," in *The Well-Being of Farm Animals,* ed. Benson and Rollin, 185–205.
35. Ibid.
36. Kristin Hagen and Donald M. Broom, "Emotional Reactions to Learning in Cattle," *Applied Animal Behaviour Science* 85, no. 3–4 (2004): 203–13.

37. Wendy K. Fulwider, Temple Grandin, D. J. Garrick, T. E. Engle, W. D. Lamm, N. L. Dalsted, and B. E. Rollin, "Influence of Free-Stall Base on Tarsal Joint Lesions and Hygiene in Dairy Cows," *Journal of Dairy Science* 90 (2007): 3559–66.
38. Grahame Coleman, "Personnel Management in Agricultural Systems," in *The Well-Being of Farm Animals*, ed. Benson and Rollin, 176–77.
39. Hemsworth, "Human-Livestock Interaction," 21.
40. M. F. Seabrook, "A Study to Determine the Influence of the Herdman's Personality on Milk Yield," *Journal of Agricultural Labour Science* 1 (1972): 45–59.
41. Temple Grandin, "The Design and Construction of Facilities for Handling Cattle," *Livestock Production Science* 49, no. 2 (1997): 103–19.
42. Temple Grandin, ed., *Livestock Handling and Transport* (Wallingford, UK: CABI International, 2007).
43. B. D. Voisinet, T. Grandin, J. D. Tatum, S. F. O'Connor, and J. J. Struthers, "Feedlot Cattle with Calm Temperaments Have Higher Average Daily Gains than Cattle with Excitable Temperaments," *Journal of Animal Science* 75, no. 4 (1997): 892–96.
44. H. M. Burrow and R. D. Dillon, "Relationships between Temperament and Growth in a Feedlot and Commercial Carcass Traits of *Bos indicus* Crossbreds," *Australian Journal of Experimental Agriculture* 37, no. 4 (1997): 407–11.

6: Pigs

1. "IQ Zoo," *Time*, February 28, 1955.
2. William Hedgepeth, *The Hog Book* (New York: Doubleday, 1998), 168.
3. Pigs have a lot of needs, so before anyone thinks about adopting a pig he should educate himself about the ways his life is going to change. There's a wonderful book entitled *The Good, Good Pig* by Sy Montgomery, about a pet pig named Christopher Hogwood, published by Random House.
4. Peter Davies, "Improving Prediction of Day of Farrowing," Proceedings of the North Carolina Healthy Hogs Seminar, http://mark.asci.ncsu.edu/HealthyHogs/book1996/book96_9.htm.
5. Timothy E. Blackwell, "Production Practices and Well-Being: Swine," in *The Well-Being of Farm Animals*, ed. Benson and Rollin, 241.
6. J. M. McFarlane, K. E. Bøe, and S. E. Curtis, "Turning and Walking by Gilts in Modified Gestation Crates," *Journal of Animal Science* 66, no. 2 (1988): 326–33.
7. Guillermo A. M. Karlen, Paul H. Hemsworth, Harold W. Gonyou, Emma Fabrega, A. David Strom, and Robert J. Smits, "The Welfare of Gestating Sows in Conventional Stalls and Large Groups on Deep Litter," *Applied Animal Behaviour Science* 105, no. 1–3 (2007): 87–101.

8. T. Grandin, S. E. Curtis, and I. A. Taylor, "Toys, Mingling and Driving Reduce Excitability in Pigs," *Journal of Animal Science* 65, Supplement 1 (1987): 230 (abstract).
9. Daniel M. Weary, Michael C. Appleby, and David Fraser, "Responses of Piglets to Early Separation from the Sow," *Applied Animal Behaviour Science* 63, no. 4 (1999): 289–300.
10. Ibid.
11. B. I. Damm, L. J. Pedersen, L. B. Jessen, S. M. Thamsborg, H. Mejer, and A. K. Ersbøll, "The Gradual Weaning Process in Outdoor Sows and Piglets in Relation to Nematode Infections," *Applied Animal Behaviour Science* 82, no. 2 (2003): 101–20.
12. Daniel M. Weary, Jennifer Jasper, and Maria J. Hötzel, "Understanding Weaning Distress," *Applied Animal Behaviour Science* 110, no. 1–2 (2007): 24–41.
13. Violaine Colson, Pierre Orgeur, Valérie Courboulay, Sébastien Dantec, Aline Foury, and Pierre Mormède, "Grouping Piglets by Sex at Weaning Reduces Aggressive Behaviour," *Applied Animal Behaviour Science* 97, no. 2–4 (2006): 152–71.
14. Blackwell, "Production Practices and Well-Being: Swine," 244.
15. Christopher A. Parratt, Kathryn J. Chapman, Christine Turner, Philip H. Jones, Michael T. Mendl, and Bevis G. Miller, "The Fighting Behavior of Piglets Mixed before and after Weaning in the Presence or Absence of a Sow," *Applied Animal Behaviour Science* 101, no. 1–2 (2006): 54–67.
16. T. Grandin and J. Bruning, "Boar Presence Reduces Fighting in Mixed Slaughter-Weight Pigs," *Applied Animal Behaviour Science* 33, no. 2–3 (1992): 273–76.
17. Blackwell, "Production Practices and Well-Being: Swine," 258.
18. The pig industry has produced its first "pig whisperer," a woman named Nancy Lidster, who has refined good pig-handling technique and produced videos and workshops. I hope her work will become better known.
19. Jon E. L. Day, Heleen A. Van de Weerd, and Sandra A. Edwards, "The Effect of Varying Lengths of Straw Bedding on the Behaviour of Growing Pigs," *Applied Animal Behaviour Science* 109, no. 2–4 (2008): 249–60.
20. Johan J. Zonderland, Maaike Wolthuis-Fillerup, Cornelis G. van Reenen, Marc B. M. Bracke, Bas Kemp, Leo A. den Hartog, and Hans A. M. Spoolder, "Prevention and Treatment of Tail Biting in Weaned Piglets," *Applied Animal Behaviour Science* 110, no. 3–4 (2008): 269–81.
21. Merete Studnitz, Margit Bak Jensen, and Lene Juul Pedersen, "Why Do Pigs Root and in What Will They Root? A Review on the Exploratory Behaviour of Pigs in Relation to Environmental Enrichment," *Applied Animal Behaviour Science* 107, no. 3–4 (2007): 183–97.
22. Heleen A. Van de Weerd, Caroline M. Docking, Jon E. L. Day, Peter J.

Avery, and Sandra A. Edwards, "A Systematic Approach towards Developing Environmental Enrichment for Pigs," *Applied Animal Behaviour Science* 84, no. 2 (2003): 101–18.

23. Amanda K. Gifford, Sylvie Cloutier, and Ruth C. Newberry, "Objects as Enrichment: Effects of Object Exposure Time and Delay Interval on Object Recognition Memory of the Domestic Pig," *Applied Animal Behaviour Science* 107, no. 3–4 (2007): 207.
24. Temple Grandin, "Effect of Rearing Environment and Environmental Enrichment on the Behavior and Neural Development of Young Pigs."
25. Suzan Dudink, Helma Simonse, Inge Marks, Francien H. de Jonge, and Berry M. Spruijt, "Announcing the Arrival of Enrichment Increases Play Behaviour and Reduces Weaning-Stress-Induced Behaviours of Piglets Directly after Weaning," *Applied Animal Behaviour Science* 101, no. 1–2 (2006): 86–101.
26. Temple Grandin, "Behavior of Slaughter Plant and Auction Employees towards Animals," *Anthrozoös* 1, no. 4 (1998): 205–13.
27. ProHand: An Interactive Computerised Training Program for Pig Stock People, PowerPoint, Slide 52, presentation at the Ohio Pork Congress, Columbus, Ohio (2006).
28. P. H. Hemsworth, G. J. Coleman, and J. L. Barnett, "Improving the Attitude and Behavior of Stockpersons towards Pigs and the Consequences on the Behavior and Reproductive Performance of Commercial Pigs," *Applied Animal Behaviour Science* 39, no. 3–4 (1994): 349–62.
29. G. J. Coleman, P. H. Hemsworth, M. Hay, and M. Cox, "Modifying Stockperson Attitudes and Behaviour towards Pigs at a Large Commercial Farm," *Applied Animal Behaviour Science* 66, no. 1–2 (2000): 11–20.
30. ProHand: An Interactive Computerised Training Program for Pig Stock People.

7: Chickens and Other Poultry

1. N. G. Gregory and L. J. Wilkins, "Broken Bones in Domestic Fowl: Handling and Processing Damage in End-of-Lay Battery Hens," *British Poultry Science* 30, no. 3 (1989): 555–62.
2. Ian J. H. Duncan, "Welfare Problems of Poultry," in *The Well-Being of Farm Animals,* ed. Benson and Rollin, 316.
3. Ibid., 302–23.
4. T. Bas Rodenburg, Hans Komen, Esther D. Ellen, Koen A. Uitdehaag, and Johan A. M. van Arendonk, "Selection Method and Early-Life History Affect Behavioural Development, Feather Pecking and Cannibalism in Laying Hens: A Review," *Applied Animal Behaviour Science* 110, no. 3–4 (2008): 2.
5. Ibid., 3.

6. This deformity has a fancy name: *tibial dyschondroplasia.* Michael Lynch, Barry H. Thorp, and Colin C. Whitehead, "Avian Tibial Dyschondroplasia as a Cause of Bone Deformity," *Avian Pathology* 21, no. 2 (1992): 275–85.
7. Steven Leeson and John D. Summers, *Commercial Poultry Nutrition*, 2nd ed. (Guelph, Ontario: University Books, 2005).
8. C. J. Savory, "Stereotyped Behavior as a Coping Strategy in Restricted-Fed Broiler Breeding Stock," in *Proceedings of the Third European Symposium on Poultry Welfare*, ed. J. M. Faure and A. D. Mills (Tours, France: French Branch of the WPSA, 1989).
9. C. J. Savory and K. Maros, "Influence of Degree of Food Restriction, Age and Time of Day on Behaviour of Broiler Breeder Chickens," *Behavioural Processes* 29, no. 3 (1993): 179–90.
10. Duncan, "Welfare Problems of Poultry," 314.
11. S. T. Millman, I. J. Duncan, and T. M. Widowski, "Male Broiler Breeder Fowl Display High Levels of Aggression toward Females," *Poultry Science* 79, no. 9 (2000): 1233–41.
12. Duncan, "Welfare Problems of Poultry," 315.
13. James Craig and William Muir, "Genetics and the Behavior of Chickens: Welfare and Productivity," in *Genetics and the Behavior of Domestic Animals*, ed. Grandin, 263–98.
14. LayWel, "Welfare Implications of Changes in Production Systems for Laying Hens; Thematic Priority: Integrating and Strengthening the ERA, Area 8.1.B.1.4, task 7," www.laywel.eu/web/pdf/deliverables%2031-33%20health.pdf.
15. Melissa Bateson, Daniel Nettle, and Gilbert Roberts, "Cues of Being Watched Enhance Cooperation in a Real-World Setting," *Biology Letters* 2, no. 3 (2006): 412–14.
16. L. S. Cordiner and C. J. Savory, "Use of Perches and Nestboxes by Laying Hens in Relation to Social Status, Based on Examination of Consistency of Ranking Orders and Frequency of Interaction," *Applied Animal Behaviour Science* 71, no. 4 (2001): 305–17.
17. B. O. Hughes, S. Wilson, M. C. Appleby, and S. F. Smith, "Comparison of Bone Volume and Strength as Measures of Skeletal Integrity in Caged Laying Hens with Access to Perches," *Research Veterinary Science* 54, no. 2 (March 1993): 202–6.
18. Tina M. Widowski and Ian J. H. Duncan, "Working for a Dust Bath: Are Hens Increasing Pleasure Rather than Relieving Suffering?" *Applied Animal Behaviour Science* 68, no. 1 (2000): 39–53.
19. Anja B. Riber and Björn Forkman, "A Note on the Behaviour of the Chicken That Receives Feather Pecks," *Applied Animal Behaviour Science* 108, no. 3–4 (2007): 337–41.
20. Tina M. McAdie, Linda J. Keeling, Harry J. Blokhuis, and R. Bryan Jones, "Reduction in Feather Pecking and Improvement of Feather Condition

with the Presentation of a String Device to Chickens," *Applied Animal Behaviour Science* 93, no. 1–2 (2005): 78.

8: Wildlife

1. Deborah Blum, "The Primatologist," review of *Jane Goodall: The Woman Who Redefined Man* by Dale Peterson, *New York Times*, November 26, 2006.
2. Dale Peterson, *Jane Goodall: The Woman Who Redefined Man* (Boston: Houghton Mifflin, 2006).
3. Mary Roach, "Almost Human," *National Geographic* (April 2008): 126–44.
4. Jill D. Pruetz and Paco Bertolani, "Savanna Chimpanzees, *Pan troglodytes verus*, Hunt with Tools," *Current Biology* 17, no. 5 (March 2007): 412–17.
5. Lori Marino, Richard C. Connor, R. Ewan Fordyce, Louis M. Herman, Patrick R. Hof, Louis Lefebvre, David Lusseau, Brenda McCowan, Esther A. Nimchinsky, Adam A. Pack, Luke Rendell, Joy S. Reidenberg, Diana Reiss, Mark D. Uhen, Estel Van der Gucht, and Hal Whitehead, "Cetaceans Have Complex Brains for Complex Cognition," *Public Library of Science Biology* 5, no. 5 (May 2007): e139.
6. Brenda McCowan, Lori Marino, Erik Vance, Leah Walke, and Diana Reiss, "Bubble Ring Play of Bottlenose Dolphins (*Tursiops truncatus*): Implications for Cognition," *Journal of Comparative Psychology* 114, no. 1 (2000): 98–106.
7. Laura Spinney, "When Chimps Outsmart Humans," *New Scientist*, no. 2555 (June 10, 2006).
8. T. Matsuzawa, M. Tomonaga, and M. Tanaka, eds., *Cognitive Development in Chimpanzees* (Tokyo: Springer, 2007).
9. C. R. Raby, D. M. Alexis, A. Dickinson, and N. S. Clayton, "Planning for the Future by Western Scrub-Jays," *Nature* 445 (2007): 919–21.
10. Giorgio Vallortigara, Allan Snyder, Gisela Kaplan, Patrick Bateson, Nicola S. Clayton, and Lesley J. Rogers, "Are Animals Autistic Savants? *Public Library of Science Biology* 6, no. 2 (2008): e42.
11. Temple Grandin, "Response to the Essay 'Are Animals Autistic Savants?'" *Public Library of Science Biology* 6, no. 2 (2008): e42.
12. Elizabeth Hess, *Nim Chimpsky: The Chimp Who Would Be Human* (New York: Bantam, 2008).
13. As I mentioned earlier, *reinforcement* is anything that follows a behavior and makes that behavior more likely to occur in the future.
14. Pryor, *Don't Shoot the Dog!*, 154.
15. Guy Gugliotta, "Rare Breed: Can Laurie Marker Help the World's Fastest Mammal Outrun Its Fate?" *Smithsonian* (March 2008).
16. T. M. Caro, *Cheetahs of the Serengeti Plains* (Chicago: University of Chicago Press, 1994).

17. Charles Siebert, "An Elephant Crackup?" *New York Times*, October 8, 2007, 40–71.
18. G. A. Bradshaw, Allan N. Schore, Janine L. Brown, Joyce H. Poole, and Cynthia J. Moss, "Elephant Breakdown," *Nature* 433 (February 24, 2005): 807.
19. David Baron, *The Beast in the Garden: A Modern Parable of Man and Nature* (New York: W. W. Norton, 2003), 6.
20. Scott M. Gende and Thomas P. Quinn, "The Fish and the Forest: Salmon-Catching Bears Fertilize Forests with the Partially Eaten Carcasses of Their Favorite Food," *Scientific American* (July 2006).
21. Allan Savory, "The Savory Grazing Method of Holistic Resource Management," *Rangelands* 5, no. 4 (1983): 155–59.
22. *Stockman Grass Farmer* (Ridgeland, MS: Valley Publishing) is an excellent source of practical information on intensive rotational grazing.
23. Emma Marris, "Plant Science: Gardens in Full Bloom," *Nature*, 440 (April 13, 2006): 860–63.
24. Richard Louv, *Last Child in the Woods: Saving Our Children from Nature-Deficit Disorder* (New York: Algonquin Books, 2008).
25. Oliver Sacks, *An Anthropologist on Mars: Seven Paradoxical Tales* (New York: Alfred A. Knopf, 1995), 121–29.
26. Arnold D. Kerr and R. Byron Pipes, "Technology: Why We Need Hands-On Engineering Education," *Technology Review* 90, no. 7 (October 1987): 38.
27. Norman L. Fortenberry, Jacquelyn F. Sullivan, Peter N. Jordan, and Daniel W. Knight, "Engineering Education Research Aids Instruction," *Science* 317 (August 31, 2007): 1175–76.
28. Temple Grandin, "Transferring Results of Behavioral Research to Industry to Improve Animal Welfare on the Farm, Ranch and the Slaughter Plant," *Applied Animal Behaviour Science* 81, no. 3 (2003): 215–28.
29. D. Fraser, "The 'New Perception' of Animal Agriculture: Legless Cows, Featherless Chickens, and a Need for Genuine Analysis," *Journal of Animal Science* 79, no. 3 (2001): 634–41.
30. Douglas H. Chadwick, "Out of the Shadows," *National Geographic* (June 2008): 106, 129.
31. Rodney Jackson and Rinchen Wangchuk, "Linking Snow Leopard Conservation and People-Wildlife Conflict Resolution: Grassroots Measures to Protect the Endangered Snow Leopard from Herder Retribution," *Endangered Species Update* 18, no. 4 (2001): 138–46. For more information on the work of the Snow Leopard Conservancy go to www.snowleopardconservancy.org.
32. Matthew J. Walpole and Chris R. Thouless, "Increasing the Value of Wildlife through Non-Consumptive Use? Deconstructing the Myths of Ecotourism and Community-Based Tourism in the Tropics," in *People and Wildlife: Con-*

flict or Coexistence?, ed. Rosie Woodroffe, Simon Thirgood, and Alan Rabinowitz (Cambridge: Cambridge University Press, 2005), 122–39.

33. Mike Norton-Griffiths, "Whose Wildlife Is It Anyway?" *New Scientist* 193, no. 2596 (March 24, 2007): 24.
34. Dietrich Dorner, *The Logic of Failure* (New York: Metropolitan Books, 1997).
35. Ibid., 95.

9: Zoos

1. Ronald R. Swaisgood and David J. Shepherdson, "Scientific Approaches to Enrichment and Stereotypies in Zoo Animals: What's Been Done and Where Should We Go Next?" *Zoo Biology* 24, no. 6 (2005): 499–518.
2. Temple Grandin, Matthew B. Rooney, Megan Phillips, Richard C. Cambre, Nancy A. Irlbeck, and Wendy Graffam, "Conditioning of Nyala (*Tragelaphus angasi*) to Blood Sampling in a Crate with Positive Reinforcement," *Zoo Biology* 14, no. 3 (1995): 261–73.
3. Megan Phillips, Temple Grandin, Wendy Graffam, Nancy A. Irlbeck, and Richard C. Cambre, "Crate Conditioning of Bongo (*Tragelaphus eurycerus*) for Veterinary and Husbandry Procedures at the Denver Zoological Gardens," *Zoo Biology* 17, no. 1 (1998): 25–32.
4. Trevor B. Poole, "Meeting a Mammal's Psychological Needs: Basic Principles," in *Second Nature: Environmental Enrichment for Captive Animals*, ed. David J. Shepherdson, Jill D. Mellen, and Michael Hutchins (Washington, DC: Smithsonian Institution, 1998), 83–84.
5. Hal Markowitz and Cheryl Aday, "Power for Captive Animals: Contingencies and Nature" in *Second Nature*, ed. Shepherdson, Mellen, and Hutchins, 48.
6. M. K. Boorer, "Some Aspects of Stereotypic Patterns of Movement Exhibited by Zoo Animals," *International Zoo Yearbook* 12 (1972): 164–68.
7. R. Clubb and S. Vickery, "Locomotory Stereotypies in Carnivores: Does Pacing Stem from Hunting, Ranging or Frustrated Escape?" in *Stereotypic Animal Behavior: Fundamentals and Applications to Welfare*, ed. G. Mason and J. Roshen (Wallingford, UK: CABI International, 2006), 58; Ros Clubb and Georgia Mason, "Animal Welfare: Captivity Effects on Wide-Ranging Carnivores," *Nature* 425 (October 2, 2003): 473–74.
8. Beat Wechsler, "Stereotypies in Polar Bears," *Zoo Biology* 10, no. 2 (1991): 177–88.
9. Markowitz and Aday, "Power for Captive Animals," 48.
10. Susan Mineka, M. Gunnar, and M. Champoux, "The Effects of Control in the Early Social and Emotional Development of Rhesus Monkeys," *Child Development* 57 (October 1986): 1241–56.

11. Joy A. Mench, "Environmental Enrichment and the Importance of Exploratory Behavior," in *Second Nature,* ed. Shepherdson, Mellen, and Hutchins, 35.
12. Rick A. Bevins and Joyce Besheer, "Novelty Reward as a Measure of Anhedonia," *Neuroscience & Biobehavioral Reviews* 29, no. 4–5 (2005): 707–14, retrieved from http://digitalcommons.unl.edu/psychfacpub/42 on March 29, 2008.
13. Martin E. P. Seligman, *Learned Optimism: How to Change Your Mind and Your Life* (New York: Free Press, 1998), 20.
14. Panksepp, *Affective Neuroscience,* 53.
15. Yudhijit Bhattachajee, "Defending Captivity," *Science* 320 (June 6, 2008): 1269. Mike Wade, "Zoos Are the Best Hope, Says Jane Goodall," *The Times,* May 20, 2008, www.timesonline.co.uk/tol/news/article3972868.ece.
16. Ibid.
17. *Conspecifics* are members of the same species.

Afterword

1. Goto, Y., and A. A. Grace, "Limbic and Cortical Information Processing in the Nucleus Accumbens," *Trends in Neuroscience* 31 (2008): 552–58.

Note to readers of *Animals in Translation*: Here is an explanation of how the core emotions compare to the behavior motivators in the Troubleshooting Guide. FEAR and RAGE are the same. PANIC is the same as sociality, and SEEKING is the same as novelty-seeking. Predatory chasing is probably motivated by SEEKING. There are some situations where biting and other aggressive behavior is motivated by FEAR.

Index

About the Authors

One of the world's most accomplished and well-known adults with autism, TEMPLE GRANDIN offers remarkable insights into animal behavior from her unique position at the intersection of autism and science. She has a Ph.D. in animal science and is a professor at Colorado State University. She is also the author of four previous books, including the classic autism memoir *Thinking in Pictures* and the best-selling *Animals in Translation.* Throughout her career Grandin has spearheaded reform of the quality of life and humaneness of death for farm animals. She lectures widely on both autism and animal science.

CATHERINE JOHNSON, Ph.D., is a writer specializing in neuropsychiatry and the brain. She cowrote *Animals in Translation* and served as a trustee of the National Alliance for Autism Research for seven years. She lives with her husband and three sons—two of whom have autism—in New York.